Lift Trucks

Lift Trucks

A Practical Guide for Buyers and Users

Daniel Bowman

Sponsored by *Modern Materials Handling* magazine

CAHNERS BOOKS
Division of Cahners Publishing Company, Inc.
89 Franklin Street, Boston, Massachusetts   02110

International Standard Book Number: 0-8436-1007-7
Library of Congress Catalog Card Number: 72-83304
Copyright © 1972 by Cahners Publishing Company, Inc.
All rights reserved.
Printed in the United States of America.
Halliday Lithograph Corporation, West Hanover, Massachusetts, U.S.A.

To
Dot, Dave and Diane Bowman
whose tolerance made this possible

Contents

	Foreword	ix
	Acknowledgments	xi
1	The Concern for Moving Materials	1
2	Terminology	6
3	Types of Lift Trucks	21
4	Lift Truck Design	40
5	Power Trains	58
6	Financing	83
7	Driver Training and Safety	96
8	Maintenance	104
9	Layout Planning	123
10	Attachments	140
11	The Unit Load, Pallets and Packaging	149
12	Storage Racks	172
13	Shipping and Receiving Docks	186
14	Yard Handling	195
15	Applications: Problem Solving with Lift Trucks	213
	Appendix – Present Worth and Capital Recovery Factors	225
	Appendix – Formulae and Conversion Factors	226
	Appendix – Lift Truck Manufacturers	227
	Index	231

Foreword

Comparison with the old-fashioned cattledrive best emphasizes the importance of materials handling. Rounding up cattle and getting them to market employ the same basic principles as the movement of goods in a modern day industrial plant.

The cattledrive involved moving a commodity from one place to another without incurring any more expense than was absolutely necessary. How closely this problem relates to the materials handler of today. Of all the handling equipment available in the modern industrial plant, probably the most familiar piece is the lift truck. Using the roundup as an analogy to materials handling, I would like to show the reader why it is so important for him to completely familiarize himself with this versatile item.

In the cattledrive, improper movement could cause a stampede or, in other words, damage the product. Today thousands of dollars are lost annually in plants throughout the country as lift trucks dump loads or run into equipment.

In that earlier era, it was the cowboy, who directed the herd. Today it is the driver or operator of the truck who rounds up the goods to be moved. However, there is great difference between the cowboy and the truck operator. The cowboy had to prove himself. He practiced good maintenance: his gun, horse, and other tools of the trade were given priority over personal comfort. In some industrial plants, anyone who can push a broom is allowed to be a truck operator.

There are many similarities between the cattledrive and materials handling in management, too. During a cattledrive, planning was always in evidence. Preplanning occurred before the drive, to take the best route to market. Followup was necessary to avoid pitfalls along the way. Each cowboy handled many cattle. In effect, each group of cattle was a unit load.

Yet with all the common sense that prevailed in that earlier day, it would appear to have been thrown out the window in many plants today. Improper loading, poor utilization of truck time, and, in some cases, no maintenance program for the trucks, take place in far too many plants across the country.

With all the grandiose schemes for plant layouts and computerized warehouses, woe to the poor man on the floor who has to operate the equipment. Far too many universities throughout the country are turning out Industrial Engineers who are theorists rather than practical men who know all phases of materials handling: engineering, equipment, personnel, and maintenance.

It is with this thought in mind that I decided to develop a book on lift trucks — a book that shows what a lift truck is, what it can and cannot do, and how to maintain it — a book that is light on theory and hard on practicality. At the end there is a directory of lift truck manufacturers. This should be of great value when the reader decides that he needs a truck.

Acknowledgments

In addition to the lift truck manufacturers listed in the back of the book, I would like to thank the following people and firms who helped make this book possible:

    Dr. Paul Eaton, Georgia Institute of Technology
    Professor James Apple, Georgia Institute of Technology
    Richard Muther of Richard Muther & Associates
    Larry Nilsen of L. B. Nilsen & Associates
    Cascade Corporation
    C & D Batteries
    Continental Motors Corp.
    Exide Power Systems Div. of ESB Inc.
    The Goodyear Tire and Rubber Co.
    Gould, Inc.
    Sperry Rand Corp.
    Uniroyal Tire Co.
    The Industrial Truck Association

Many others have contributed as well to provide background material for this book. Unfortunately, many of the notes taken in the field were not attributed to any particular person or place, and to those persons or companies not mentioned, my most grateful appreciation. An additional word of thanks for all the technical brochures, catalogs, and other technical data that the manufacturers of materials handling companies have so generously provided.

Lift Trucks

1 | The Concern for Moving Materials

Trying to trace the history of the lift truck may not seem important to the average plant manager or engineer. His main concern is to solve today's problem and, as they say, "Let's get on with it."

Because of the truck's great usage during World War II, probably most people think the lift truck was developed in the 1940s. The widespread use of lift trucks and other handling devices during the War makes it easy to understand this belief. Yet, the date for the introduction of the first mechanical lift truck is placed around 1890 and the first self-propelled truck appeared in 1917.

World War II proved the value of good materials handling methods. It might be debatable as to who had the best equipment or armies, but no one would question who had the best production and supply lines. Through efficient materials handling methods, the Allied Armies had no trouble in being supplied with arms and war materials.

Not only were many handling devices developed around the turn of the century, but also many of the ideas and much of the methodology. Frederick Taylor, the Gilbreths, and others were promoting many of the handling concepts we now take for granted; as industrial engineers, they had proven that man was more efficient if he had mechanical aids to help him. Through war and peacetime, speed was essential, resulting in a need for equipment to lift, carry, and place items. As the need for space became a premium, the industrialist found that growing up was far more practical than growing out. The end result was the birth of the "cube" and, without the lift truck, the story wouldn't be possible.

Several companies have been in the industrial truck business for more than 50 years. Yet, new lift truck companies have also been ap-

2 | Lift Trucks

pearing since the War. In 1950, the Taylor Machine Works started building industrial lift trucks and, just a few years ago, Drexel Dynamics Corp. started building lift trucks.

How did companies begin building lift trucks? There are some 40 different firms making lift trucks and each one has its own story. However, almost all of them got into the business to help someone solve a materials handling problem.

Clark Equipment Co. went into the business because it had its own handling problem at its Buchanan, Mich., plant in 1917 (fig. 1-1).

Fig. 1-1. A pioneering start by the Clark Equipment Co. is this 1917 industrial truck they built for their own use in Buchanan, Mich. Courtesy of Clark Equipment Co.

The solution was to build a platform truck. As could be predicted, the truck's popularity caught on with the company's other plants and with other manufacturing companies. So they started building them and have been in the business ever since.

When the Hyster Co. was founded, its first products were handling devices for the Northwest logging industry. The same was true of the Taylor Machine Works, except that they built logging equipment for the Southern logging industry. In both cases, it was an evolutionary process, and one handling device led to another and it wasn't long before they were building lift trucks.

Once a materials handling problem is solved for one person, it is inevitable that someone else will have the same problem. Naturally, that person is going to approach a company which has the solution to its own handling problem, regardless of whether or not that company is in the materials handling business. In fact, many new handling developments have originated in plants that did not manufacture materials handling equipment.

Materials handling is a strange business as it is a function that adds nothing of value to a product. Yet, materials handling equipment sales are over a billion dollars a year. Just how can a function that does not change a product's shape, weight, color or any other of the product's characteristics be so important?

The Big Change

The materials handling function does change something — something that is so evident that hardly anyone can tell you what it is. The main reason for moving materials is to change their usefulness. Groceries stored in a warehouse are of no value to a hungry person, unless he has a way to get them from storage to his dining room table. Coal in the ground is of little value, but if it is transported to a power plant, the coal can become useful in generating power for thousands of persons.

Materials movement can be measured in many ways: speed, volume, time of arrival, etc. Usually the primary function of moving materials is based on the following:

Time: When do you want them? Should they be here today, tomorrow, or a week from now?

Distance: How far do the materials have to be moved? Distance will greatly affect the mode of transportation used.

Quantity: How many do you need? A dozen or, perhaps, millions?

Rate: Just how fast do you want to receive the materials? One per minute, one per hour, or one per day?

The Universal Manipulator

Fortunately, the plant engineer seldom makes a mistake in buying a lift truck. The main problem is the degree of efficiency required in solving the particular problem. A lift truck can be obtained in almost any size, any capacity, and with enough attachments to move almost any object. In fact, the lift truck is the universal manipulator.

However, in order to find an economical solution to your materials handling problem, a systematic analysis should be made of the handling problem. This is the main thought behind this book, not to make a truck designer out of you, or a mechanic, or an operator, but rather to help you become an efficient user of handling equipment and, in particular, the lift truck.

Look at your present operation; take out your notebook and actually walk through your facility. Look at the workers. Are they moving materials as efficiently as possible? Is there a worker standing idle, waiting for materials? Are the aisles clear, can men and machines get through?

How about the materials handling equipment that you do have: are the lift trucks running? If they are, do any of them have oil leaks? Take a walk to the maintenance department: is there a truck being serviced or is the truck just sitting there, waiting to be repaired?

Materials handling is big business. Your plant spends thousands of dollars on lift trucks and other handling items each year and the figure will increase. Nevertheless, the reason all that equipment is necessary is that operations become more efficient and profitable as a result of using good methods. The savings in manhours, prevention of accidents, and the efficient utilization of equipment and facilities will be many times the original investment of the handling equipment.

As we said, "Let's get on with it," let's find out about the lift truck (fig. 1-2).

FIG. 1-2. As they say, "You have come a long way, baby." This is a recent model introduced by Eaton Corp., Industrial Truck Div. Courtesy of Eaton Corp., Industrial Truck Div.

2 | Terminology

The cry of the vendor at a ballgame is, "You can't tell the players without a program." This statement holds true for lift trucks — with a slight variation. It's difficult to talk about lift trucks without knowing the terminology. Materials handlers have developed a language of their own and this chapter covers the basic terminology of the lift truck.

In general, a forklift truck is a vehicle designed to pick up, carry, and stack unit loads of supplies and equipment. Although there are two major power sources for lift trucks, differences between the two are very few, except the power plants.

Basically, the lift truck consists of a power plant, the truck body, and a mast system for raising and lowering loads. The biggest difference in trucks, other than power plants, is whether the vehicle is counterbalanced or noncounterbalanced.

Counterbalanced Trucks

The counterbalanced truck lifts loads outside the framework of the truck and is the best-known model. Most of the weight of the counterbalanced truck is carried on the front wheels which, in essence, act as a fulcrum. The weight center line of the truck is at such a distance from the front wheels that its moment is greater than the center of gravity of the load plus its distance from the front wheels.

In summary, the moments of the load and forks at the front end of the truck must never be greater than the moments or counterbalancing effect of the truck, body, and power plant. In figuring the capacity

Terminology | 7

the truck must never be greater than the moments or counterbalancfactor.

The load capacity of the vehicle is the maximum load at a "stated load center" that the truck can transport and/or stack to a specified height.

The load center is the horizontal distance between the front vertical engaging face of the forks to the center line of gravity of the load. Therefore, if a truck is rated 2,000 pounds at 24 inches, it means that it will carry a 2,000-pound load as long as the load center of gravity is no more than 24 inches from the face of the truck. The drawings in figures 2-1 and 2-2 illustrate the load center of a counterbalancing truck, and the chart (fig. 2-3) shows how load capacity is rated.

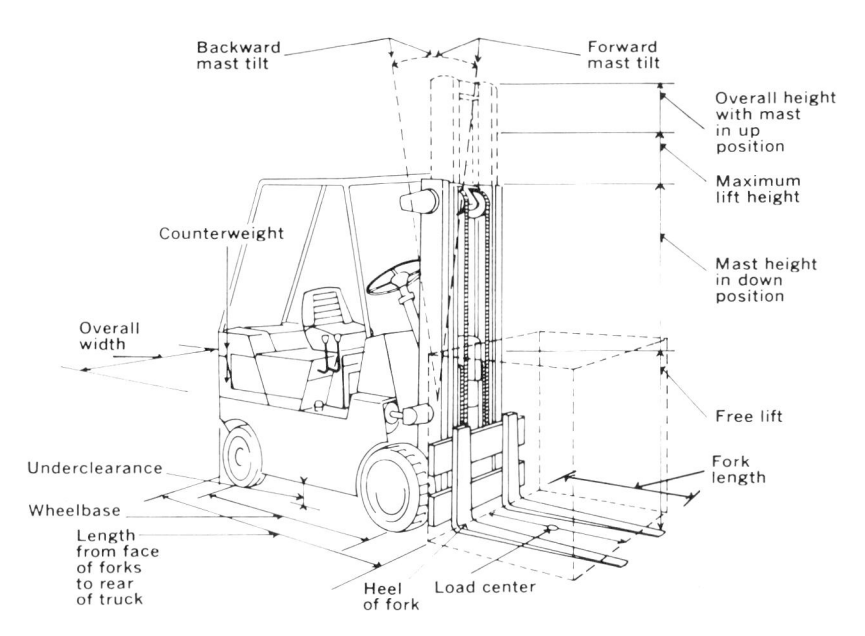

FIG. 2-1. Lift truck terminology. Courtesy of White Materials Handling Division, White Motor Corporation.

Noncounterbalanced Trucks

Noncounterbalanced trucks carry the load within the framework of the truck. Examples would be the straddle truck and the outrigger

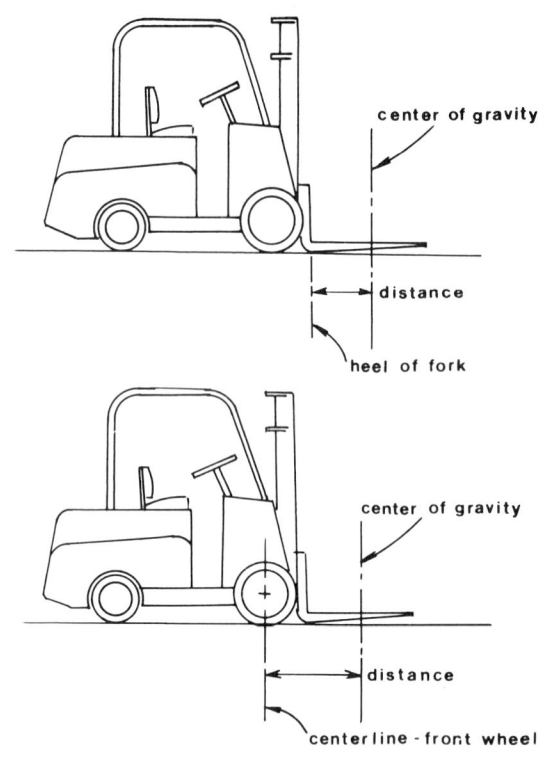

FIG. 2-2. The load center is figured from the heel of the fork to the center of gravity of the load, while the moment is figured from the load's center of gravity to the center line of the front wheel. Because of the moment effect, the load capacity will decrease as the load center increases as shown in the chart, figure 2-3. Conversely, the capacity increases as the load center decreases until the mechanical capacity of the lifting mechanism is reached (in this case, 9,000 lbs).

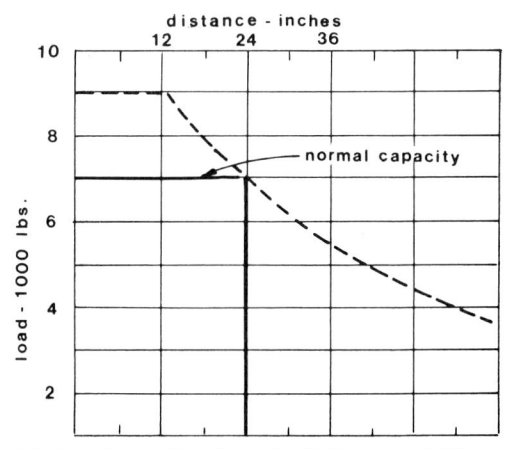

FIG. 2-3. Load capacity chart for 7,000-pound lift truck.

truck. The load-carrying ability of these vehicles is dependent entirely on the mechanical strength of the truck's mechanisms.

In some cases, such as a sideloading lift truck, the load is picked up as though by a counterbalanced truck, placed on a platform on the truck body, and then transported as though by a noncounterbalancing truck.

Lifting Characteristics

The most important feature of a lift truck is, naturally, the lifting mechanism. Loads are lifted by an elevating carriage attached to the mast of the truck (fig. 2-4). The forks (tines), which are attached to the carriage, secure the load.

When lift trucks were first introduced, they had fixed masts. This meant that the mast of the truck had to be as high as the distance that the load was to be lifted. When it was found that in some applications the trucks would be traveling in low-ceiling areas and yet

Fig. 2-4. Carriage assembly on the mast of the lift truck. Backrest guard prevents loads from falling backward off the forks when carriage is elevated. Courtesy of Towmotor

have the need for higher stacking, the telescoping mast was developed.

The telescoping mast allows a lift truck to maneuver high lifts and at the same time have a low profile when carrying a load. Each section of the telescoping mast is called a stage. Therefore, a two-stage mast will have two sections, one inside the other. Lift height involves the following two terms: *Maximum Lift Height,* the measurement from the floor to the load carrying surface adjacent to the heel of the forks when the forks are elevated to their highest position (fig. 2-5); and *Free Lift,* which is a term used with telescoping masts and is the distance from the floor to the surface of the forks when the forks are in their highest position without the mast telescoping, as in figure 2-6.

Most masts have the ability to tilt in addition to raising or lowerings loads. The greatest advantage of tilting is that it makes the load more secure. By tilting the load backwards, the load would have to travel up the forks in order to slip off if the truck made a sudden stop. Tilting forward allows a truck to slip under and release loads more easily.

In some instances, such as loading materials on a beverage truck, the tilting ability allows loads to be placed on a slanted shelf. Many of the smaller walking-type trucks have no tilting ability. The larger walkie/rider trucks (trucks where the operator can either walk along and control the truck through a handle and yet mount the truck and ride it) will usually have a backward tilt of about 6 degrees — mainly for securing the load. Most rider-type trucks have a forward tilt slightly less than the backward tilt (5 to 6 degrees forward and 10 to 12 degrees backward). The larger yard handling trucks will have a greater tilting capability (15 degrees forward and 15 degrees backward), since the truck may be on a grade and the load has to be placed on a level surface, such as happens in construction projects (fig. 2-7).

Truck Body

Most industrial trucks are constructed with a unitized body which consists of the chassis and body being welded together into one solid piece. The body encompasses the power train, hydraulic system, steering mechanism, and wheels.

Counterbalanced trucks employ a lifting mechanism attached to the

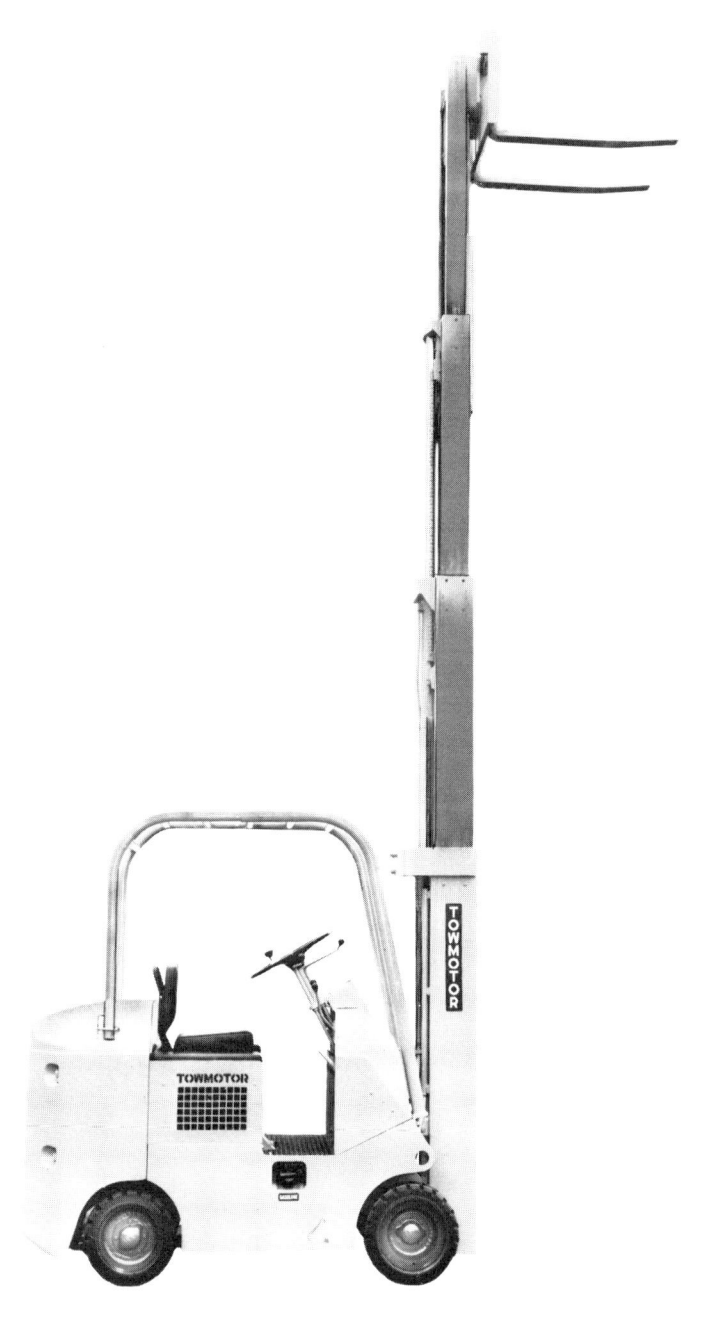

FIG. 2-5. Four-stage mast allows loads to be lifted to heights of 17 ft 6 in., yet in a retracted position requires a height of only 75 in. Courtesy of Towmotor

12 | Lift Trucks

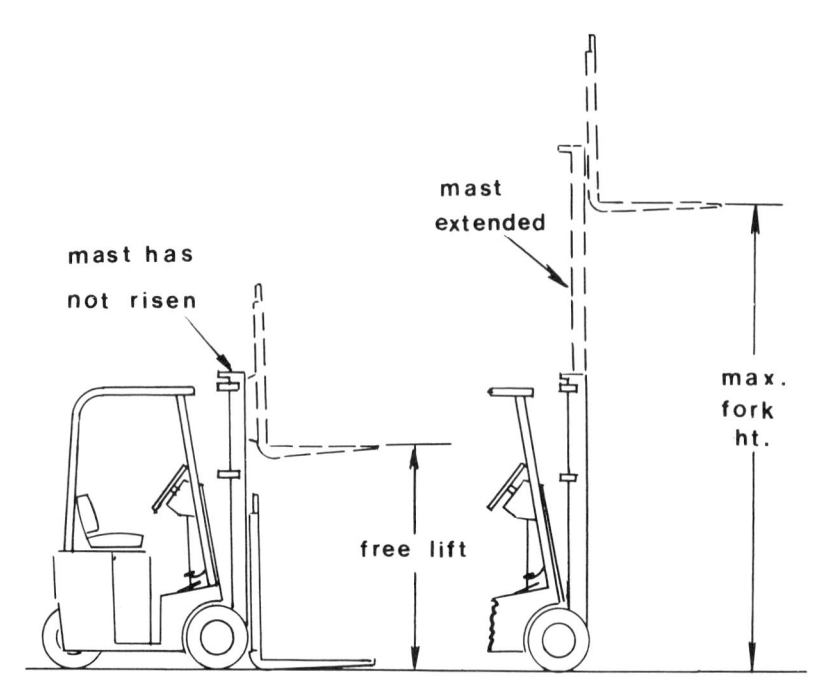

FIG. 2-6. Free lift is the height to which the forks can lift without elevating the mast.

front end and either the battery weight, as in the case of electric trucks, or a heavy casting, as in the case of gas trucks, supplies the additional weight needed for the counterbalancing effect. In gas trucks the counterweight serves as added protection for the radiator (fig. 2-8).

An important relationship to the truck's body and power plant is its gradability (grade climbing ability). Lift trucks are often required to climb ramps or dock plates and, in addition, they must also be able to clear the grade at the top (fig. 2-9).

Grade is usually given as a percentage and is calculated by dividing the rise in feet by the horizontal distance of the grade (fig. 2-10). If the grade is given in degrees or needs to be expressed in degrees, tables 2-1 and 2-2 supply figures for conversion. The grade clearance

Terminology | 13

FIG. 2-7. Tilt is the amount of movement off a vertical line the mast can make either forward or backward. Usually forward tilt is less than backward tilt. Courtesy of Hyster Company

14 | Lift Trucks

Fig. 2-8. A counterweight is added to the back of the truck to counteract the load placed on the front of the truck. The truck is really acting as a lever with the fulcrum being the center line of the front wheels. Courtesy of Hyster Company

of the truck is the maximum grade change that the truck will clear without any part of the underside of the truck touching the floor.

The underclearance of the truck is the distance between the lowest point of the underside of the truck and a level floor. The wheelbase is the distance from the center line of the front wheels to the center line of the back wheels. As a result, grade clearance will be determined by underclearance and wheelbase (fig. 2-11).

Gradability of electric trucks is about 15%; gas trucks, 20% to 25%. Some gas trucks have a higher gradability but would have more application in construction sites.

Although most lift trucks were not designed to pull loads, some have that capability and the truck should have a drawbar pull rating. Drawbar pull is figured on the basis of the towing operations being per-

Terminology | 15

FIG. 2-9. Gradability is the truck's ability to climb a ramp. Courtesy of Towmotor

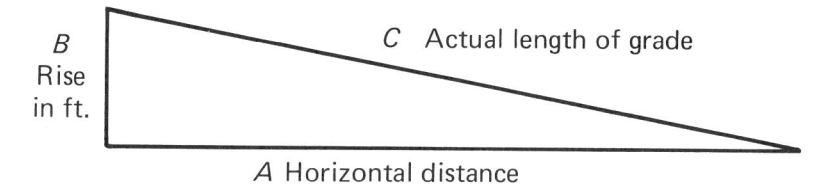

B Rise in ft.
C Actual length of grade
A Horizontal distance

Formula for figuring percentage of grade:

$$\frac{100 \times B}{A} = \text{grade in \%}$$

Example: B = 5 ft. and A = 50 ft.

$$\frac{100 \times 5}{50} = 10\%$$

Since the horizontal distance may be difficult to measure, the length of the grade C may be substituted for A for reasonable accuracy of grades up to 20%.

FIG. 2-10. How to calculate the grade percentage.

16 | Lift Trucks

Table 2-1. Conversion of grade percentages into degrees

Percent	Degrees	Percent	Degrees	Percent	Degrees
1	0°34'	11	6°17'	21	11°52'
2	1° 9'	12	6°51'	22	12°24'
3	1°43'	13	7°25'	23	12°57'
4	2°18'	14	7°58'	24	13°30'
5	2°52'	15	8°32'	25	14° 2'
6	3°26'	16	9° 5'	26	14°34'
7	4°	17	9°39'	27	15° 7'
8	4°34'	18	10°12'	28	15°39'
9	5° 9'	19	10°45'		
10	5°43'	20	11°19'		

Table 2-2. Conversion of degrees into grade percentages

Degrees	Percent	Degrees	Percent	Degrees	Percent
1	1.75	6	10.51	11	19.44
2	3.49	7	12.28	12	21.26
3	5.24	8	14.05	13	23.09
4	6.99	9	15.84	14	24.93
5	8.75	10	17.63	15	26.80

formed on smooth dry concrete or comparable surface having a coefficient rating of 0.90. To calculate the necessary power to move a given load, a figure of 20 pounds for each 1,000 pounds pulled is used. Therefore, if a truck were to pull 20,000 lbs, it would have to have a drawbar pull rating of (20,000/1,000) × 20, or 400 lbs. Where a grade is involved, refer to table 2-3 for calculating the required drawbar pull.

Loss of efficiency occurs when surfaces are wet or icy. To obtain the actual drawbar pull rating in instances of adverse surface conditions, multiply the rating obtained from table 2-3 by the appropriate corrective factor listed below:

concrete, wet	0.66	ice at 0° F, with chains	0.45
asphalt, wet	0.66	ice at 32° F	0.04
hard-packed snow	0.22	ice at 32° F, with chains	0.22
ice at 0° F	0.09		

Another term associated with tipping characteristics of the truck is its stability. Previously tipping was mentioned in relation to the load

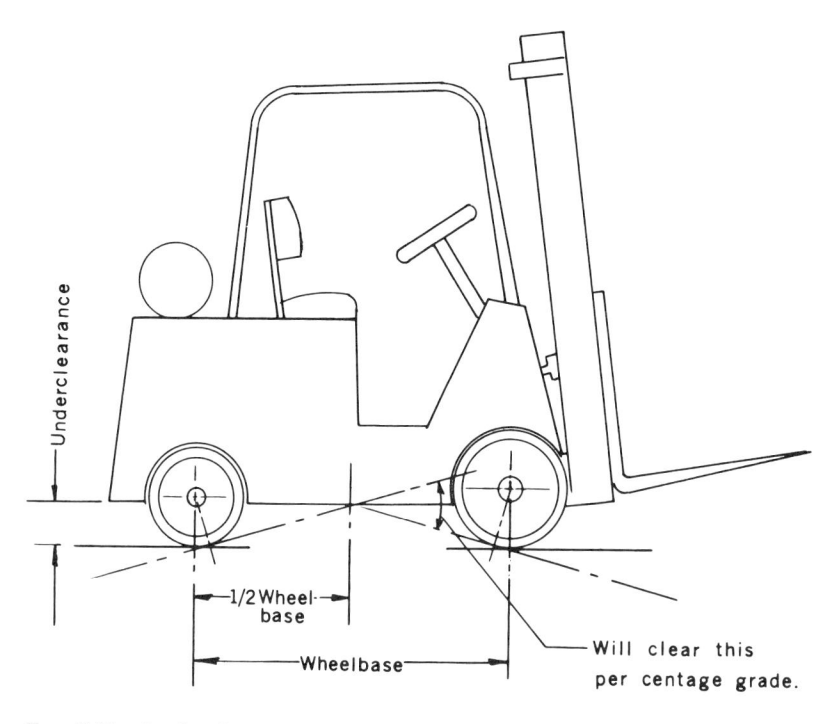

Fig. 2-11. Grade clearance depends on the truck's underclearance and its wheelbase.

center of the truck and causing the truck to tip forward. The stability of a truck is its resistance to overturning on its side. Stability would come into play as the truck turned a corner — if the truck were traveling too fast, it would tip over.

A standard feature on most modern lift trucks is an overhead guard. The guard protects the driver of the truck from any falling loads when the forks are in an elevated position (fig. 2-4, 2-5, 2-7, 2-8, 2-12).

Turning Radius

One of the most important features of any lift truck is its turning radius. Turning radius is important, as this feature determines the minimum width of an aisle in which the truck can stack (fig. 2-13). The outside turning radius determines the aisle width and is the radius of the largest circle formed by the outermost projection of

18 | Lift Trucks

Table 2-3. Drawbar pull required for towing a load

Load (lbs)	Percentage of grade						
	0%	1%	2%	4%	6%	8%	10%
1,000	20	30	40	60	80	100	120
2,000	40	60	80	120	160	200	240
3,000	60	90	120	180	240	300	360
4,000	80	120	160	240	320	400	480
5,000	100	150	200	300	400	500	600
6,000	120	180	240	360	480	600	720
7,000	140	210	280	420	560	700	840
8,000	160	240	320	480	640	800	960
9,000	180	270	360	540	720	900	1,080
10,000	200	300	400	600	800	1,000	1,200
20,000	400	600	800	1,200	1,600	2,000	2,400
30,000	600	900	1,200	1,800	2,400	3,000	3,600
50,000	1,000	1,500	2,000	3,000	4,000	5,000	6,000
100,000	2,000	3,000	4,000	6,000	8,000	10,000	12,000

Fig. 2-12. This overhead guard retracts, allowing the truck to enter areas of low headroom. Courtesy of Clark Equipment Co.

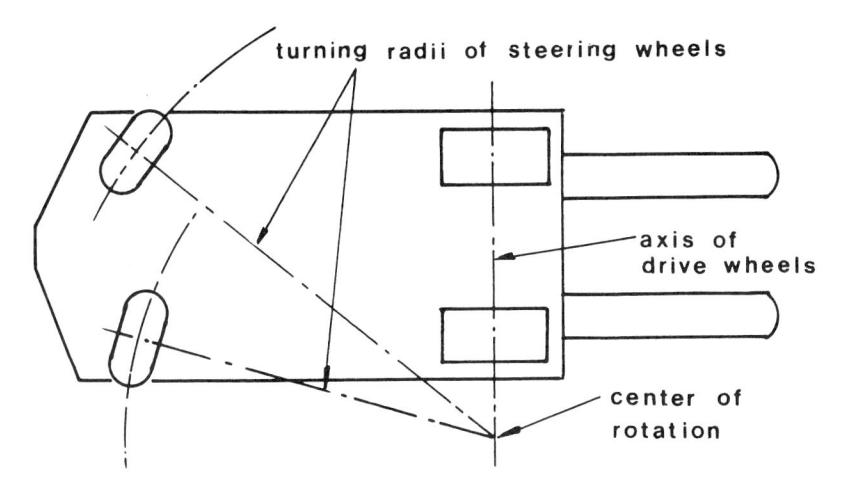

Fig. 2-13. A truck's turning radius is important, as it determines the amount of aisle space required for maneuvering the truck.

the truck, with the steering mechanism at the optimum steering angle.

The inside turning radius helps establish operation clearances (such as in intersecting aisles) and is the distance from the wheels' pivot point to the nearest inside projection of the truck.

As long as the load being carried is confined within the limits of the forks, there is no effect on the turning radius. However, if the load projects forward or over the sides, it is best to lay out the operation on paper and calculate the proper turning radius.

Other Terminology

In terms of an automobile, the speed of a lift truck isn't exactly impressive. However, speed does have a relation to the truck's productivity and two speeds are listed for a truck.

The first, the truck's travel speed, is given in mph and defines how fast the truck travels horizontally. The other is called a lift speed and is the vertical travel speed of the lifting device (forks) in fpm (feet per min).

So far we have been speaking of the truck only, but there are also a few terms the reader should know that are associated with the truck's functions. The first term is *tiering,* which is the process of placing one load on top of another. If the truck can stack two tiers, it means the two loads can be stacked, one on top of the other.

If loads are to be placed in rows, then there must be a method of placing them in storage and a method of getting them out of storage. One method is referred to as FIFO, which means the first load placed in storage is the first load taken out of storage. The reverse is called LIFO, which means the last load in is the first load out.

Another method of storage is called *random storage.* This term usually refers to racked storage and means that loads are not placed into storage under any sequential pattern.

These are the basic terms that one will encounter when using or specifying lift trucks. Other terminology may be limited to specific styles of trucks or equipment, such as the terms for various style racks and their employment. Such terms will be covered under each specific chapter as the need arises.

Bibliography

Haynes, D. Oliphant. 1957. *Materials handling equipment.* Philadelphia, Pa.: Chilton Co.
1965. Industrial engineering terminology manual. *Journal of Industrial Engineering* Nov–Dec.
Bowman, Dan. 1968. Tow cart systems. Part 2, Tow tractors. *Plant Eng.* June 13
White Mobilift. *Common sense truckology manual.* Hopkins, Minn.: White Industrial Div., White Motor Corp.

3 | Types of Lift Trucks

When trying to describe basic lift truck styles, one encounters a seemingly countless variety of units from which to choose. Trucks can be classified as riders, walkies, pneumatics, electrics, narrow aisles, etc., etc.; in general, the name explains the mode of operation.

Nevertheless, the easiest place to start is with the truck without any forks — the tow tractor. Although not really a lift truck, a tow tractor is part of any plant's truck fleet and is usually loaded or unloaded by lift trucks.

Tow Tractors

Tow tractors can be gas or electric powered and can be classified as either a walkie, rider type, or automatically controlled unit. A tow tractor has the advantage of one power source moving a great many loads, whereas a lift truck usually moves just one load at a time. Tractors can traverse nonfixed routes and are rated by their drawbar pull.

Another point to check when selecting a towing tractor is its braking capability. Power to get the trailer rolling is one thing, but stopping it can be another, especially if the truck is going down a ramp. High speeds and large loads, as in an airport, require dual braking systems. Lighter industrial tow tractors usually have only one set of brakes with a "deadman's" control. If distances are short, less than 200 ft, then a walkie tractor is used.

Equally important when choosing a tow system is the selection of trailers. The turning radius of the trailers may be much greater than

22 | Lift Trucks

that of the tow tractor (fig. 3-1). Five styles of trailers are available (fig. 3-2), and the platform style is dependent upon the load being carried. In some cases, a load dumping mechanism is employed to speed the unloading process.

When tractor routes are repetitious and fairly fixed, an automatically guided tractor can become most economical. The guided tractor can be easily programmed for almost any route and has the flexibility of being manually controlled whenever needed (fig. 3-3).

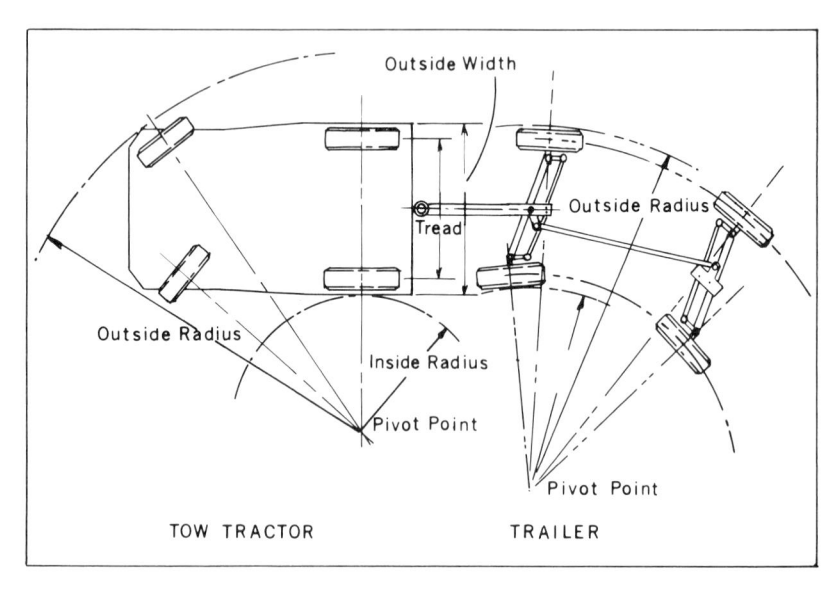

FIG. 3-1. Turning radii of a tractor versus that of a four-wheel knuckle steering trailer. Pivot point is intersection of the main wheel axis and the turning radii of the steering wheels.

An optical guidance system is the simplest and most economical of the three types of guidance systems. By using a line (either painted or pressure-sensitive tape) that contrasts in color with the floor, a light-sensing unit in the guidance system causes the steering wheel to follow the line. The towpath must be kept clean, as any paper or other object that obscures the guideline can cause the unit to stop. Optical guidance is most useful when routes have to be changed periodically.

Types of Lift Trucks | 23

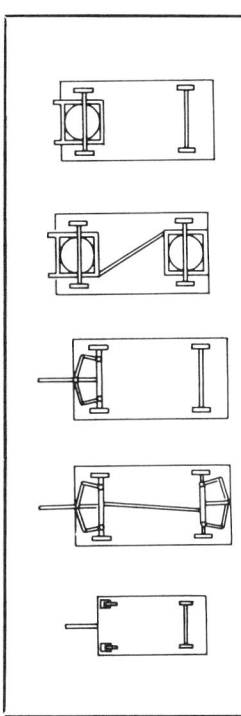

1. **Fifth-wheel** steering (single carriage turnable steering)—The front axle is fitted to a forecarriage pivoting around a king pin. Widely used for heavy duty trailers where turning restrictions are not critical.

2. **Dual-end fifth-wheel** steer (double carriage turnable steering)—Employs a fifth wheel at each end of the trailer. This method provides a sharp turning radius in relation to over-all dimensions. The steering of each carriage is interconnected and usually one can be locked out for single-end steering.

3. **Two-wheel knuckle** steer (single carriage Ackermann steering)—Front wheels are on stub axles pivoting around king-pins, the rear wheels being on a fixed axle.

4. **Four-wheel knuckle** steer (double carriage Ackermann steering)—All wheels are on stub axles pivoting around king pins with the steering of both pairs of wheels interconnected. Allows the most perfect tracking of vehicles in train and assures utmost stability in sharp turns.

5. **Caster** steering—Two rear wheels are fixed while the two front wheels are free castering. The running gear provides maximum maneuverability and the wheels track instantly in any direction.

Fig. 3-2. Five different styles of trailer steering mechanisms. Each one tracks differently, which factor will have to be taken into account when laying out aisles. Courtesy of Clark Equipment Co.

The second type of system (magnetic) follows an energized guidewire placed in the floor. Magnetic systems are for more permanent installations and allow switching and multiple tractor usage on the same route. Tractor speed is about 3 mph and the unit is equipped with a highly sensitive bumper that will cause the unit to stop immediately if the bumper comes into contact with an object. Optical and magnetic systems are shown in fig. 3-4.

The third system is radio control. This method would seldom be used, as an operator would be necessary. However, radio control may be used in conjunction with the other two methods.

Some guided tractors feature a sonic detector to prevent the unit from bumping into objects. Through the use of a solid state sonic detector, the tractor slows down if an object is within 15 ft of the front

24 | Lift Trucks

FIG. 3-3. Automatic controlled tractor trains follow either a line embedded in the floor or painted on the floor. Courtesy of Eaton Corp., Industrial Truck Div.

a.

b.

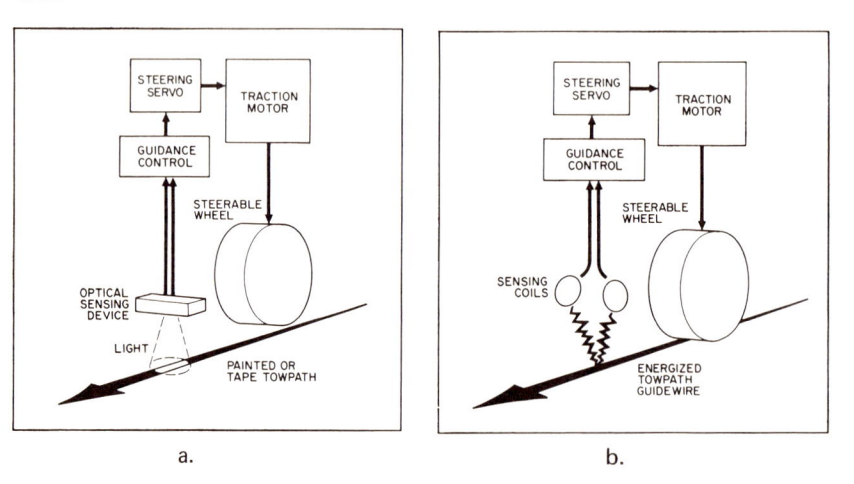

FIG. 3-4. The two guidance systems used in automatic tractors. The optical system as shown in (a) is the simplest and most economical. However, the path must be clean, for if the guideline is obscured, the tractor will stop. The magnetic system (b) is more permanent and allows switching.

of the truck and the unit comes to a complete stop within 5 ft. Other safety features are a flashing light and horn.

Platform and Pallet Trucks

Pallet and platform trucks are vehicles designed to transport loads. A platform truck is just what the name implies — a moving platform on which a load can be placed. Many varieties are available, such as gas or electric powered, stand-ups, etc. Their chief disadvantage is that they are either manually loaded or depend upon some external handling device to load them.

The pallet truck has the same function as a platform truck except that it has the capability of loading itself (fig. 3-5). The pallet truck has two elevating forks that can be inserted below the top deck of a

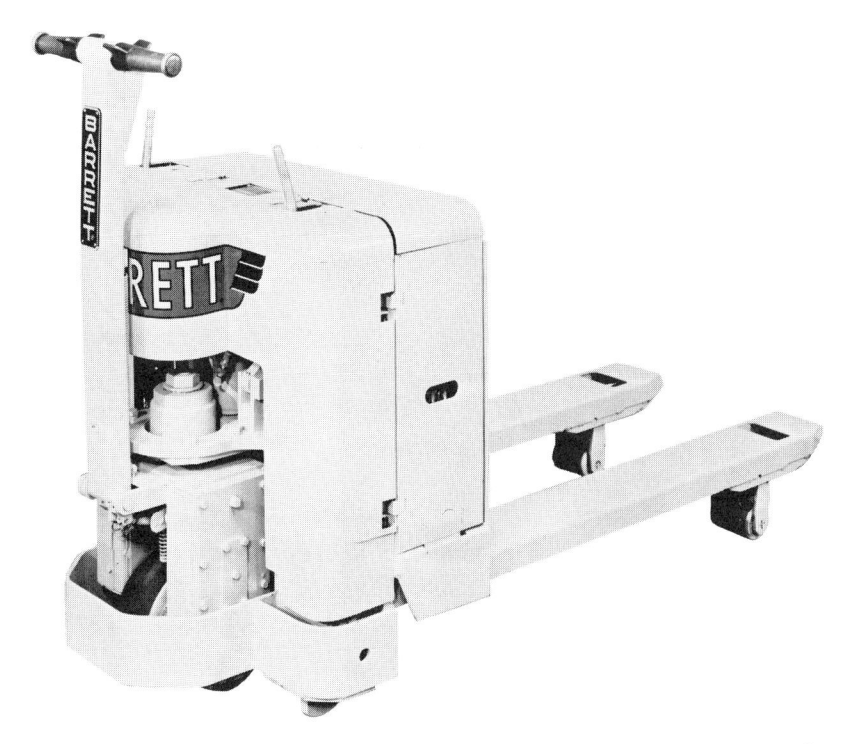

Fig. 3-5. Battery-powered pallet truck is used for transporting and not for stacking. Courtesy of Barrett-Cravens

26 | Lift Trucks

pallet. Wheels are located at the end of the forks and are articulated so that when the load is raised off the ground, the wheels extend downward to help support the load.

Some pallets have a bottom deck and, as a result, some pallet trucks have several sets of wheels on the ends of the forks. They are articulated and spring-loaded so that the load-carrying wheels retract and the other set of wheels ride over the bottom deck of the pallet (fig. 3-6).

A variation of the pallet truck is the "stillage" or skid truck. Instead of forks, it has a platform that can raise the skid or "stillage" off the ground. (See chapter 11 for the difference between a skid and pallet.)

Often a pallet or skid truck will have a rider's platform for long-distance travel. Also some pallet trucks have a swing-down platform that can be lowered onto the forks so it can operate as a skid truck.

Skid and pallet trucks solve the problem of handling occasional loads. Although not a tiering truck, some do have the capability of making small lifts. Since most states limit the weights that a man or

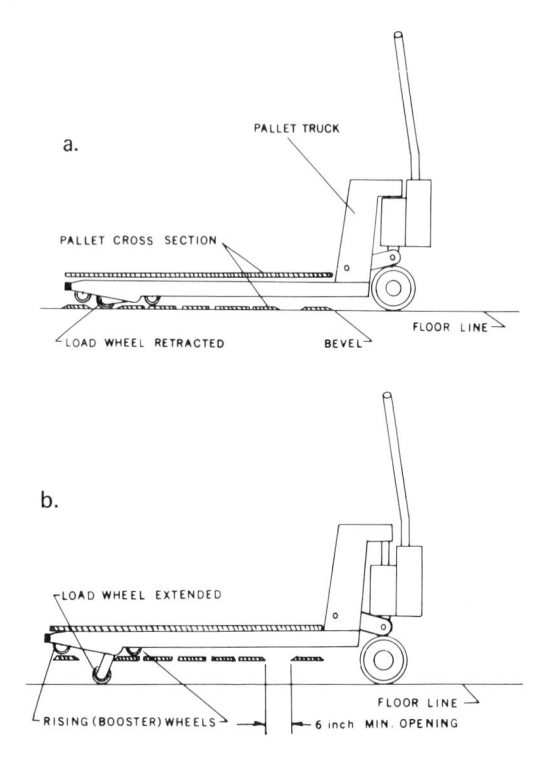

Fig. 3-6. When pallet trucks insert their forks into a pallet, the load wheels retract into the tines (*a*). Wheels extend through opening (*b*) to raise pallet.

woman can lift, almost all manufacturing plants use these trucks. There are many manually operated versions of these units which use mechanical or hydraulic lifting mechanisms. Slower in operation than a powered unit, they nonetheless greatly increase a worker's ability to perform handling operations as well as reduce the danger of a worker straining himself through manual lifting. The time lost and medical costs incurred by one worker injuring himself possibly could cost many times the price of this type of equipment.

Counterbalanced Trucks

The counterbalanced truck is described in the second chapter. However, it comes in many varieties, either gas-powered or electric. There also is a wide range of fork attachments and lifting heights available.

Some units are equipped with three wheels instead of four to provide a very tight turning radius. Others employ pneumatic tires for use on rough terrain. Drivers can be seated or, in some cases, stand up when driving the vehicle. The stand-up model (fig. 3-7) is used when the operator has to get on and off frequently. Walkie versions would be used where travel distances are short and there is infrequent usage.

Basically used for stacking operations, fork trucks are also used for short hauls up to 500 ft. One advantage of the counterbalanced truck is that the load is completely free of any objects in front of the truck (as contrasted with the narrow aisle truck with its outriggers).

Narrow Aisle Trucks

One disadvantage of the counterbalanced truck is that it requires a fairly wide aisle (10 to 12 ft) in which to maneuver. With the recent emphasis on taking advantage of the "cube" (namely, conserving space) in materials handling planning, more and more narrow aisle trucks are appearing. Operating in aisles as narrow as 6 ft, the unit has outriggers straddling the load.

The forks on a narrow aisle truck are located between two outriggers or straddle arms. The outriggers have wheels on the end; therefore, they support the load along with the drive wheels. This contrasts with the counterbalanced truck where most of the weight is placed on the drive wheels.

A wing-type pallet may be used to allow the truck to pick up loads

28 | Lift Trucks

Fig. 3-7. Stand-up counterbalanced truck is used when operator has to leave the truck frequently.

wider than the truck's outriggers. Contrasted with the counterbalancing truck, the outrigger unit's load capability is dependent upon the mechanical strength of the fork lifting mechanism.

The first rack tier should be raised above the height of the outrigger when this type of truck is used in rack operations. An outrigger-type is extremely maneuverable and, since no counterweight is required, the truck is very light when compared to a counterbalanced truck. This makes it ideal for operating in elevators or on floors with a low floor loading capability as in older mill-type buildings.

Reach Trucks

Reach trucks come in two styles; both act as counterbalancing trucks when the load is extended in front of the truck's front wheels. When carrying loads, both styles act as outrigger trucks and carry the load within the wheelbase of the truck. However, they differ in the mechanism to extend the load.

One style of reach truck has the mast moving forward and backward on the outriggers. The difficulty with this truck is that the mast prevents the outriggers from slipping under a storage rack when a load is being extended.

A reaching carriage-type truck, however, has forks that move forward on a pantograph mechanism when engaging a load. By extending the truck's reaching capability, loads can be stacked two deep, thus saving the space needed for an aisle at the back of a double storage rack (fig. 3-8).

Another advantage of a reach truck over other narrow aisle trucks is that, by extending the forks (whether by the mast moving forward or through the pantograph mechanism), loads can be picked up in front of the outriggers. This operation eliminates the need for loads narrower than the outriggers or the use of wing-type pallets.

The second style, a more recent development, has a side-loading carriage similar to that of a storage retrieval machine. The unit follows a line embedded in the floor (similar to that of the automatically guided tractor) and can store or retrieve loads automatically (fig. 3-9).

Sideloaders

The sideloader was an American innovation, but found greater usage in Europe before finally gaining popularity in the U.S. Sideloaders are not as maneuverable as standard lift trucks, but they really don't have to be. The sideloader travels forward down a storage aisle with the forks facing perpendicular to the racks. This allows the truck

30 | Lift Trucks

Fig. 3-8. Pantograph mechanism allows forks to extend beyond the ends of the outriggers. With the forks extended, truck then acts as a counterbalanced truck. Courtesy of Clark Equipment Co.

to stop in front of a rack location and to position loads directly into and out of the rack structure.

Most sideloaders are equipped with the mast in the center of the truck. The mast moves back and forth laterally across the bed of the

Types of Lift Trucks | 31

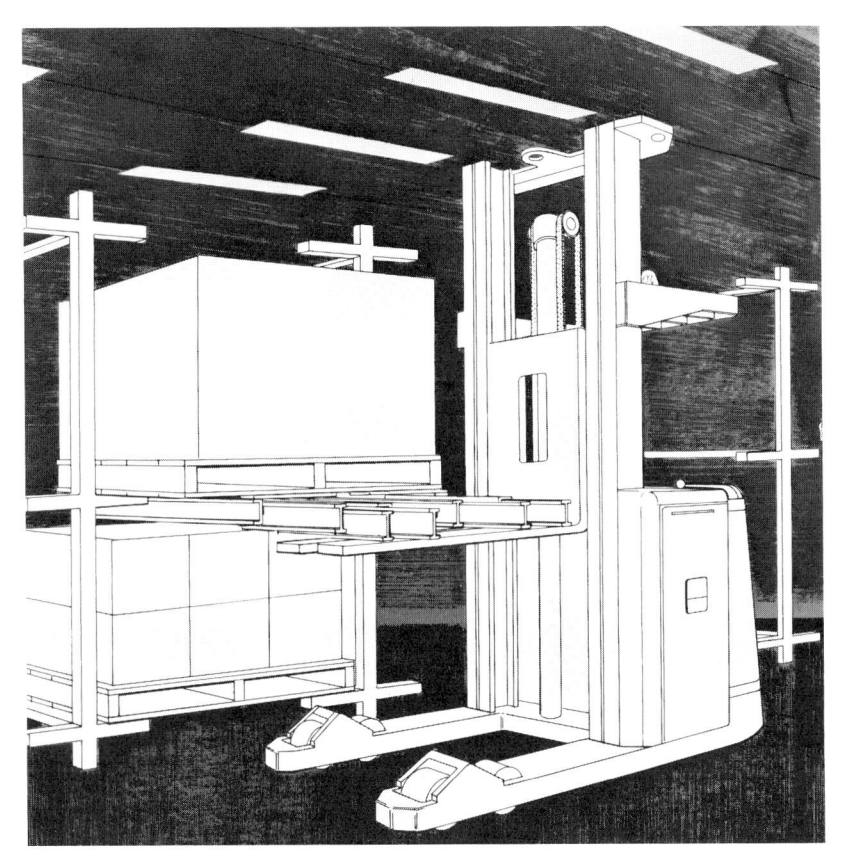

FIG. 3-9. Automatically guided outrigger truck with side-loading carriage has many of the advantages of a smaller stacker system, yet the economies of a lift truck. Courtesy of Eaton Corp. Industrial Truck Div.

truck. The chief advantage of the sideloader is that extremely long loads, such as steel pipe and lumber, can be carried without any danger of spilling the load (fig. 3-10). In addition, the load travels parallel to the aisle, thus taking up less aisle space than the standard lift truck.

A British truck manufacturer has a side-loading model available that has omnidirectional capability. All four wheels are swivel mounted, allowing the truck to act as a front loading vehicle when its wheels are rotated 90 degrees. It makes a wide front-loading lift truck, but this feature greatly enhances the truck's maneuverability.

32 | Lift Trucks

Fig. 3-10. A sideloader can operate in narrower aisles than a standard lift truck. The unit also has a bed to place long loads for a more stable ride. Courtesy of Allis-Chalmers

The main disadvantage of the sideloader is that it operates from one side only. Therefore, if the unit deposits a load on one side of an aisle, it must go to the end of the rack structure and turn around before it can remove or place a load in the other side of the rack.

Another disadvantage of the sideloader is similar to that of the outrigger truck. In narrow aisles, the unit cannot store loads any lower than the height of the truck's bed.

Straddle Carriers

The straddle carrier (fig. 3-11), as the name implies, straddles the load to be lifted. It can carry loads no higher than the inside height of the truck and cannot store in rack structures. One advantage is that the vehicle can carry long loads securely over rough surfaces. Originally

FIG. 3-11. A straddle truck is primarily a transport vehicle. Some attain speeds of over 50 mph. Courtesy of Hyster Company

designed to carry lumber, its use has spread to other industries that require a truck to carry long loads in outside storage.

Loads for a straddle truck are elevated off the ground by blocks of wood or bolsters. The bolster, similar to a wing-shaped pallet, allows the load lifting shoes to come under the load. Primarily a transport vehicle, straddle carriers usually have higher speeds than other handling devices (some can go more than 50 mph).

Not originally designed for stacking, recent versions are built extra high so that the truck can place one load on top of another. For tighter

34 | Lift Trucks

FIG. 3-12. One of the newer style lift trucks has a swinging front end, allowing it to act as a sideloader or a standard counterbalanced truck. Courtesy of British Aircraft Corp.

Types of Lift Trucks | 35

turns, some straddle trucks have four-wheel steering. The truck, designed for large loads (some have capacities of 50,000 lbs), is usually a team vehicle working in conjunction with another handling vehicle such as a fork truck or yard crane.

Swiveling Mast Trucks

A truck that combines the best features of the sideloader and the standard lift truck is the swiveling mast unit. A rather recent innovation, the truck can carry a load down an aisle in a manner similar to that of a lift truck, stop at its proper location, swivel the mast, and place the load in a rack without requiring an extra wide aisle (fig. 3-12). Several variations of the swiveling mast have appeared, including the truck with an articulating front end (fig. 3-13) and a special unit with a multidirectional fork on the end of a boom (fig. 3-14).

Any special lift truck, such as the swiveling mast unit, will be at a premium price and could even require a longer waiting period for de-

Fig. 3-13. This particular truck has an articulated mast. The operator rides with the load and has excellent visibility. Courtesy of Drexel Industries, Inc.

36 | Lift Trucks

Fig. 3-14. Another special type is this truck with the forks mounted on the end of a boom. The truck has an extremely low profile. Courtesy of Standard Mfg. Co.

livery. Nevertheless the efficiency gained may be worth the price and the wait.

Order-picking Trucks

The standard industrial lift truck made the unit load practical. In addition, its ability to tier or stack materials made possible the maximum utilization of space. Needless to say, broken lots or less than pallet loads were difficult to handle. By placing materials on a pallet, small loads could be easily handled by lift trucks, but all available space was not being utilized. Goods could be placed in storage racks through the use of a pallet, but unless one wanted to place several items on one pallet and then place the pallet in and out of the storage rack when-

Types of Lift Trucks | 37

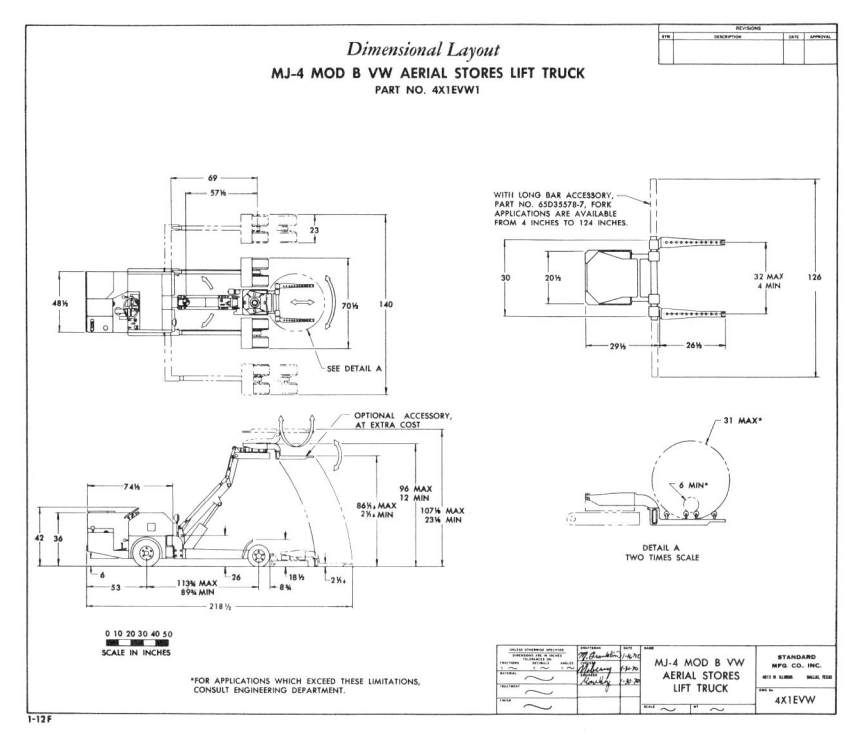

Dimensional layout for fig. 3-14.

ever an item was needed, less than pallet loads were still not profitable to handle.

A recent innovation, the order-picking truck, changed all this (fig. 3-15). By placing an operator's stand on the forks of a narrow aisle truck, an operator could drive a truck in an aisle at any elevation and pick individual items out of storage — at any level — without first having to return to the ground.

Many order-picking trucks have wheels on their sides and are guided by rails located at the base of the storage racks. This allows the driver to raise and lower himself without worrying about guiding the truck. When finished loading in one aisle, the driver can then steer the truck himself to the next aisle or take the load to a shipping dock or a point of usage.

38 | Lift Trucks

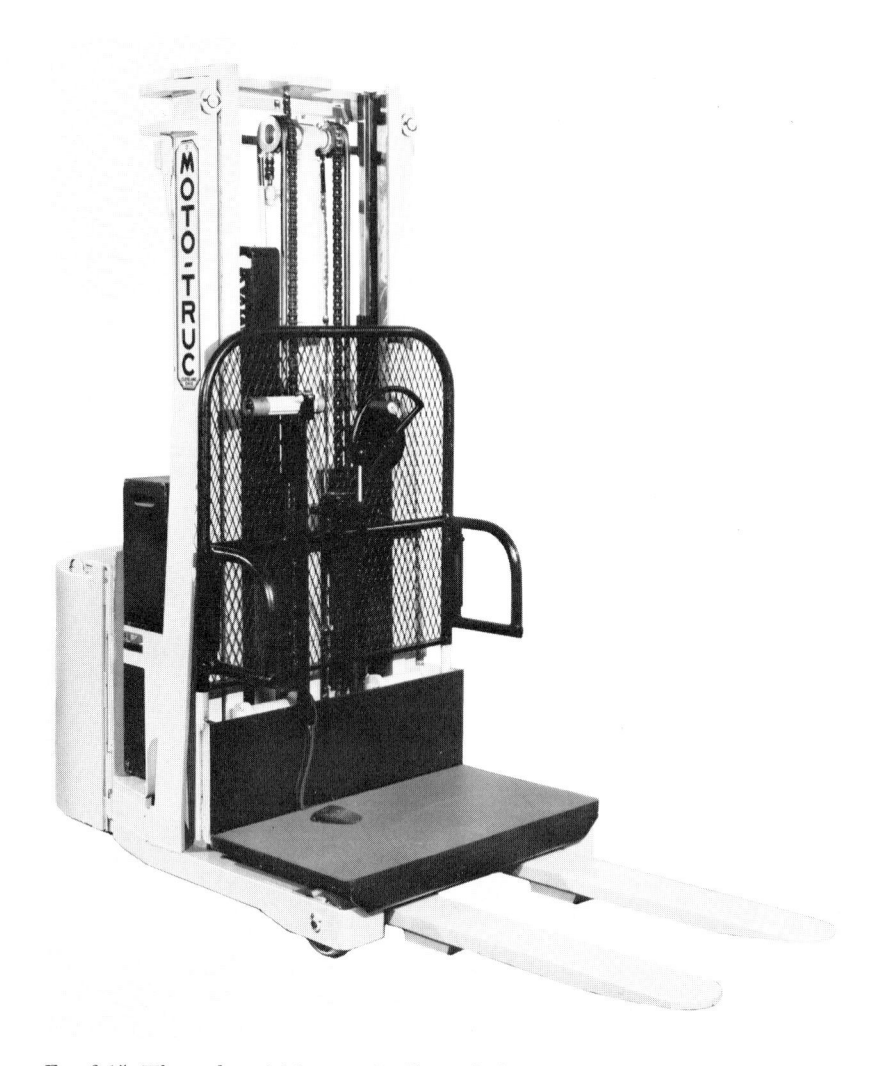

Fig. 3-15. The order-picking truck allows tiering of mixed or broken case loads in racks. Operator controls the truck from the platform and can drive the unit at any height he desires. Courtesy Otis Material Handling, Otis Elevator Company.

Many order-picking trucks are constructed with the operator's platform so that it can be removed and the truck operated as a standard narrow aisle vehicle. A recent innovation has an order-picking truck combined with the guidance control of the automatic guided tractor, allowing the truck to follow a wire buried in the floor. The truck can store up to 24 ft in height, making the unit quite competitive with smaller storage-retrieval systems.

As long as one wants to pay the price, almost any kind of lift truck can be obtained.

Summary

As long as a mast is attached to a vehicle, you have a lift truck. The fact that the lift truck is such a versatile item means that many unusual combinations can be developed. One such example is a mast attached to an over-the-road truck tractor. This allows the truck to pull a trailer over long distances, the trailer to be detached and unloaded like a lift truck.

This flexibility is probably what has made the lift truck so popular through the years. It wasn't so long ago when it was thought that the lift truck had reached its zenith. However, with continuous improvements in power plants, hydraulic systems, attachments, speeds and operating flexibility, the versatile lift truck will be around for a long time to come.

4 | Lift Truck Design

At last we reach the heart of the truck. Just what are the various components and how do they work? Included are the lifting mechanism, the hydraulic system, and steering and control.

The selection of a lift truck is not just a matter of price. Each lift truck has its own special set of features and the truck should be rated against other trucks according to the desires and needs of the user.

A simple way to compare one manufacturer's truck against another is to set up a rating system. For simplicity's sake, we will use a point system of ranking the trucks. Areas to be analyzed are: Performance (25%), Handling characteristics (15%), Power plant (15%), Safety features (15%), Maintainability (15%), Price (15%). These are arbitrary characteristics, as each purchaser must decide for himself what is important and determine how much weight should be placed on each area of analysis.

Every truck is rated in each category according to a point system: 5, excellent: 4, very good; 3, good; 2, fair; 1, poor; and 0, not satisfactory. As a result, a method of rating three different manufacturers may look something like table 4-1, in which the number of points given a truck in each category is multiplied by the percentage value assigned to the category. The results, when added, give a basis for comparison.

In table 4-1, manufacturer C's lift truck has the highest rating. However, manufacturer B's truck isn't too far behind and, with its better price, it could be the truck to choose. Yet, if the truck will have extensive use and if production is quite dependent upon this truck, then it would be better to pay the higher price and get the best performance lift truck.

Table 4-1. Analysis system for selection of lift truck

Analysis factor	Manufacturer A Points	Pts × %	Manufacturer B Points	Pts × %	Manufacturer C Points	Pts × %
Performance (25%)	5	1.25	3	0.75	4	1.00
Handling characteristics (15%)	4	0.60	3	0.45	4	0.60
Power plant (15%)	4	0.60	4	0.60	5	0.75
Safety features (15%)	2	0.30	5	0.75	4	0.60
Maintainability (15%)	1	0.15	4	0.60	4	0.60
Price (15%)	4	0.60	5	0.75	3	0.45
Totals		3.50		3.90		4.00

Actually each area in the chart can be broken down further with a rating made on characteristics in each. If several trucks are being selected, it is advantageous to make as careful an analysis as possible to insure the selection of the best truck.

Design Features

A lift truck is made up of many components (fig. 4-1), and one should understand the mechanics of the truck when buying a new one. It is advisable to take the plant's most competent driver along when selecting a truck. He can put each truck through its paces and, by using the rating system described above (he should rate only the handling characteristics area), he will feel that he has had his say in selecting the truck. Likewise, a maintenance foreman can also express his opinion on the maintainability of the truck.

Hydraulic Systems

Outside of the power plant (which is covered in the next chapter), few subsystems in the truck are as important as the hydraulic system. The hydraulic system powers the lifting mechanism of the mast and the

42 | Lift Trucks

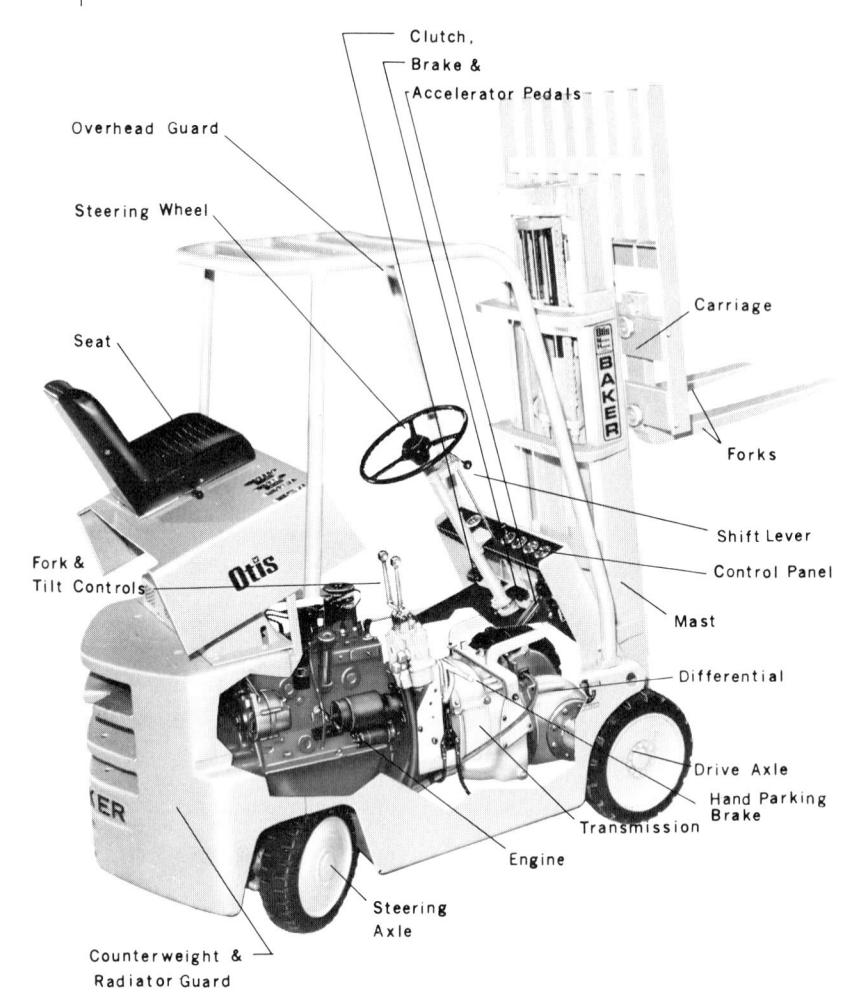

FIG. 4-1. The various components of a counterbalanced lift truck. Courtesy Otis Material Handling, Otis Elevator Company

power steering. In some cases it is used for power brakes and the transmission.

A typical hydraulic circuit consists of a reservoir, pump, control valves, lift (raise) cylinder, and tilt cylinders (fig. 4-2). The oil flows in a loop from the reservoir to the pump, through the control valves, and back to the reservoir. When the operator wants to tilt the mast or raise the forks, he moves the appropriate valve spool. Depending upon

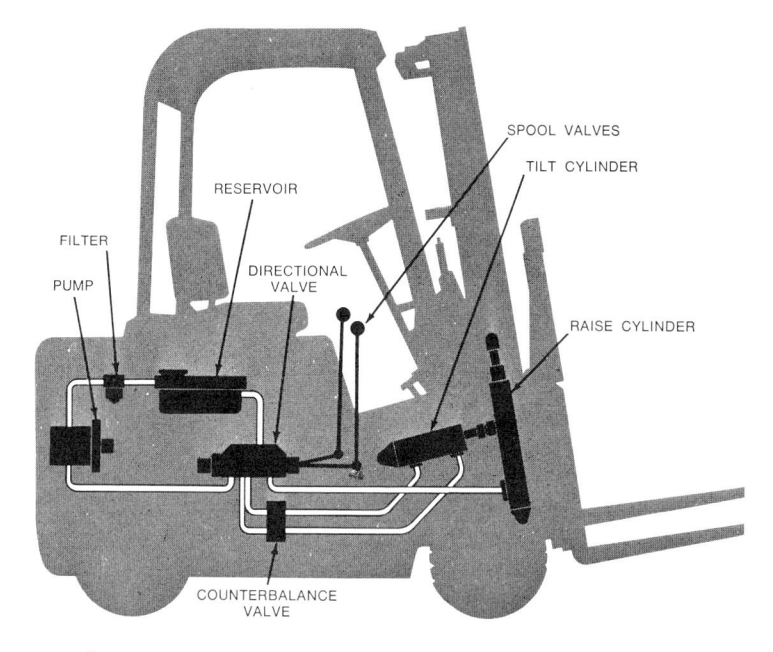

Lift Truck Design | 43

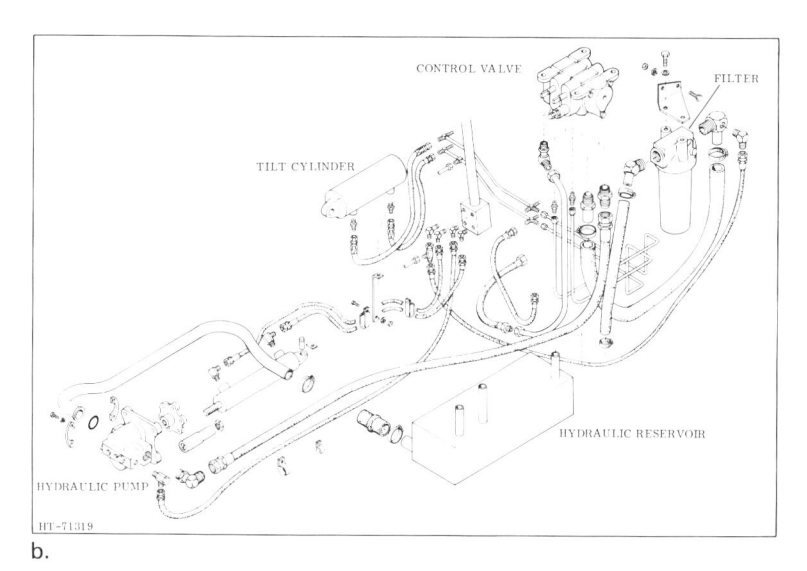

FIG. 4-2. Phantom view of lift truck shows the location of the components of the hydraulic system (a), while schematic drawing (b) shows the oil flow. Courtesy of Sperry Rand Corporation and Allis-Chalmers.

44 | Lift Trucks

the movement desired, the valve causes part of the oil to continue bypassing the valve and the remainder of the oil is admitted into the tilt or mast cylinder. Smoothness of the control circuit is an important operating characteristic as it prevents jerking of the forks when raising a load. Probably the best way to determine this characteristic is to actually test the truck under load.

Most lift trucks use either a gear or vane type pump (fig. 4-3). The

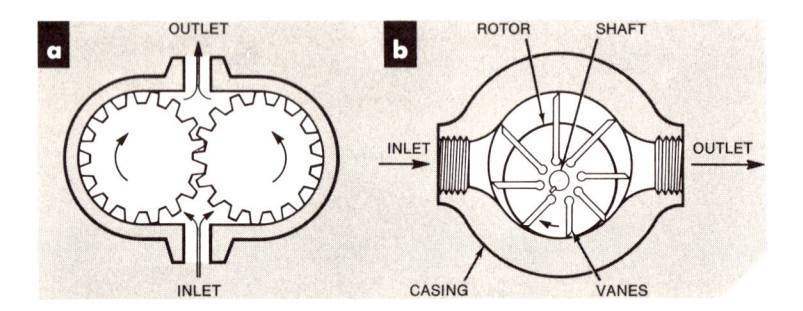

FIG. 4-3. The two main types of pumps used are gear and vane pumps. Pumps should be accessible for ease of maintenance. Courtesy of Sperry Rand Corporation.

pump may be directly driven by the engine or by V-belts. The main requirement of the pump location is its accessibility for maintenance.

The reservoir serves several purposes including: filtering out contaminants; separating the air out of the oil; and acting as a heat exchanger to dissipate the heat out of the oil. Naturally, the larger the reservoir, the better its heat transfer capability. The reservoir should also be exposed to as much outside airflow as possible to help its cooling capability. In some trucks the tank is placed above the pump to prevent any possibility of the pump cavitating.

Two types of hydraulic cylinders are used. One is a single-acting cylinder that raises and lowers the forks; the other is a double-acting cylinder that tilts the mast.

Two types of power steering are also used in lift trucks, both optional. When power steering is used, it will also be part of the hydraulic system (fig. 4-4). The steering cylinder is double acting and operates similarly to the tilt cylinder. Movement of the steering wheel operates a control valve which allows oil to flow into one side of the steering cylinder or the other, depending upon in which direction the operator wants to go. In this particular case there is no mechanical linkage between the steering wheel and the steering axle.

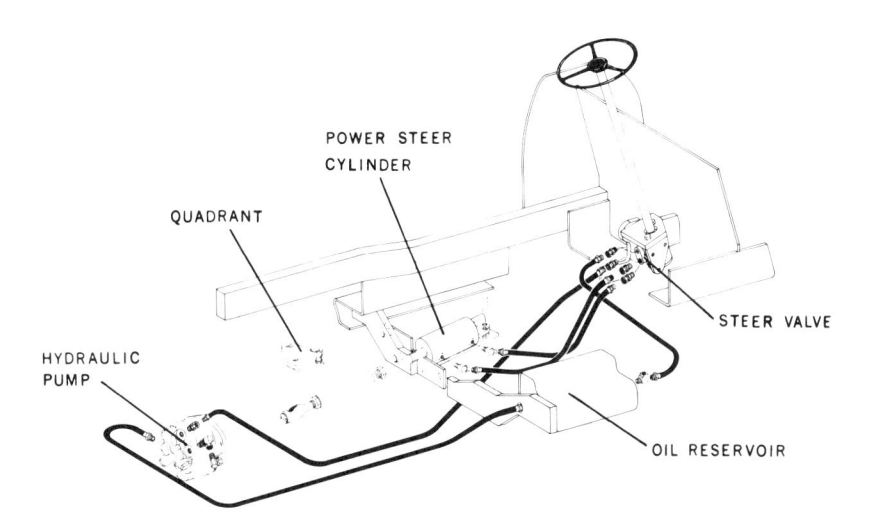

FIG. 4-4. The steering cylinder and valve are mounted separately and have no mechanical linkage between them. Courtesy of Allis-Chalmers.

However, some power steering units have a mechanical linkage connecting the steering wheel and axle in addition to the hydraulic system (fig. 4-5). This provides a failsafe capability and, in the case of a power failure, the steering reverts to manual control.

Power steering is almost a "must" on all large pneumatic tired trucks — usually starting at 3,000 lbs — and at 6,000 lbs and up for solid tired trucks. However, if the truck is being used constantly, power steering will help reduce driver fatigue regardless of the truck's size.

Mast selection is important. The mast consists of steel channels that ride inside one another, with the carriage riding on the inside channels (fig. 4-6a). Carriage rollers hold the carriage inside the channels and allow the carriage to move up and down.

The mast is raised and lowered by the hydraulic cylinder. Attached

46 | Lift Trucks

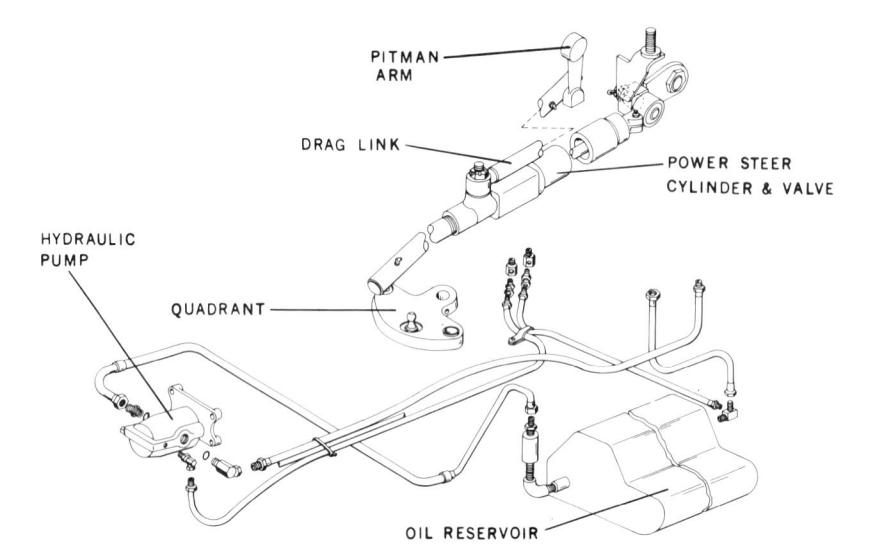

FIG. 4-5. In a power steering unit, the cylinder acts as a booster to mechanical linkage. Courtesy of Allis-Chalmers.

to the top of the hydraulic cylinder is a piece called the yoke. On each side of the yoke is an idler chain sprocket. A roller chain is placed over each sprocket and one end is affixed to the truck by way of the mast or cylinder. The other end of the chain is attached to the carriage. Thus, when the cylinder is raised, in addition to lifting the mast, it indirectly lifts the carriage through the roller chain. "Free lift" is the amount of movement the carriage can make before the cylinder yoke starts lifting the mast (fig. 4-6*b*). A standard upright usually has just enough free lift to raise the load off the floor so the truck can go through a doorway without the top of the mast striking the door.

A free-lift upright allows a load to be lifted half the total height without extending the uprights. Other combinations of channels and cylinders allow 3 and 4 stage uprights (fig. 4-6*c* and 4-6*d*). There are two types of bearings used to guide the mast in its telescoping action: bronze pads or anti-friction bearings.

The forks or front end attachments are mounted on the carriage (fig. 4-7). Several mountings are possible with forks. The most common is the hook-mounted fork. However, there are other available types such as a swing fork (fig. 4-8), which can swing upward or chisel

Lift Truck Design | 47

STANDARD UPRIGHTS
a.

FREE-LIFT UPRIGHTS
b.

3-STAGE UPRIGHTS
c.

4-STAGE UPRIGHTS
d.

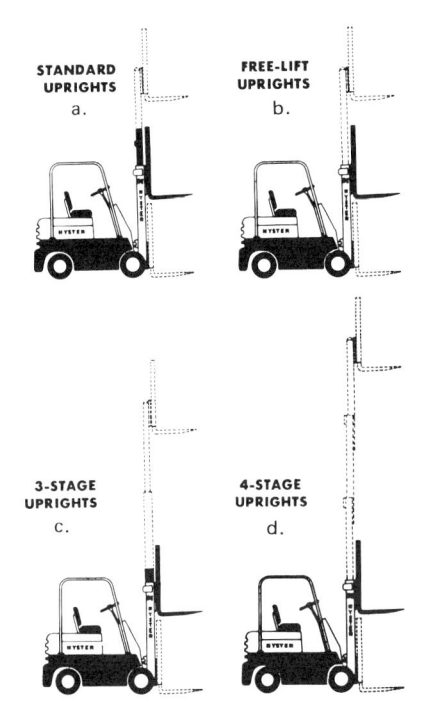

Fig. 4-6. By staging the uprights, the truck's profile in the lowered position remains the same, while the lifting height is greatly increased. Free-lift is shown in *b*. Courtesy of Hyster Company.

forks, which can work with either mounting. Forks can be obtained in any length and width. Normally forks are 4 to 6 inches wide. The most critical stress spot of the fork is at its heel. Therefore, one should use the thickest fork possible for an operation.

Steering

Previously, we mentioned that the steering for a truck can be powered or mechanical. However, in order for the wheels to turn there has to be a steering axle. Several types are available, but the one most frequently used is the articulated axle that is quite similar to a car axle (fig. 4-9).

The steering axle is pivoted in the center so it can ride over bumps and yet not subject the truck chassis to torsional stresses. This axle is the only one that has springs or is cushioned; the drive axle has no springs.

Some trucks have a tricycle-type gear which allows a very tight turn-

48 | Lift Trucks

Fig. 4-7. The forks are mounted on the carriage. This carriage has a backguard, which prevents loads from falling off when the mast is tilted backward. Courtesy of Eaton Corp., Industrial Truck Div.

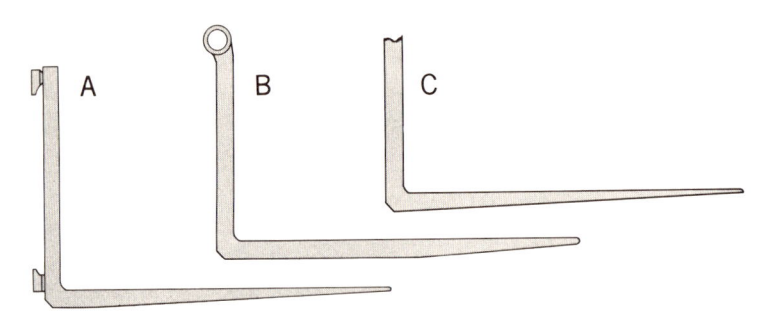

Fig. 4-8. Fork styles are shown in this illustration. Standard hook-on fork is A; swing-up fork, B; and chisel fork, C.

Lift Truck Design | 49

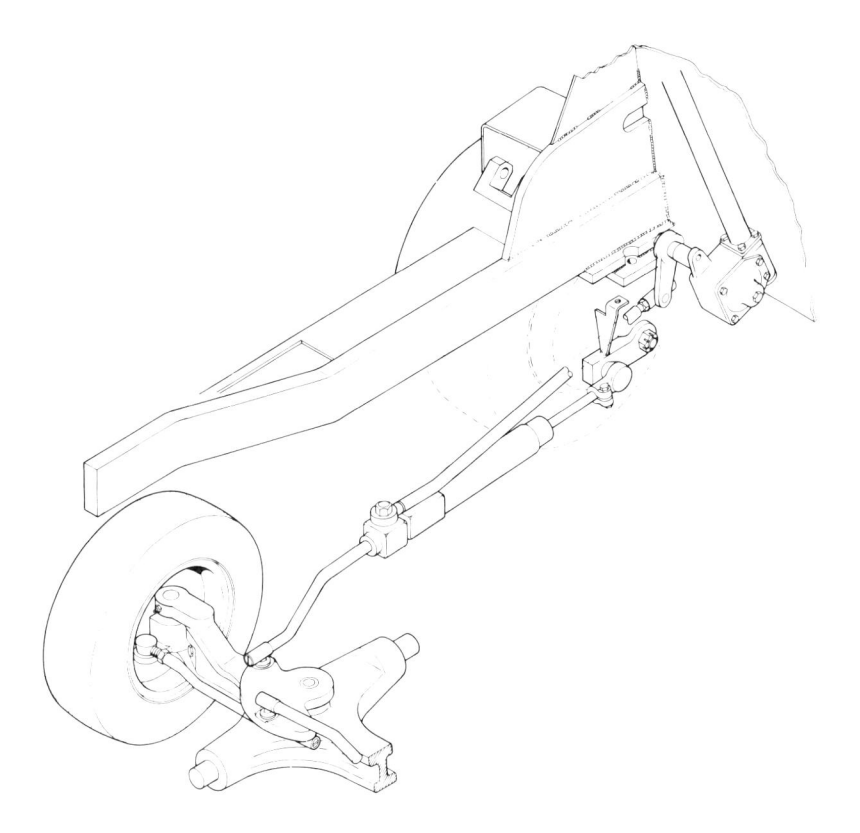

Fig. 4-9. Articulated steering axle is quite similar to that of an automobile. Courtesy of Allis-Chalmers.

ing radius. On small electric trucks the drive motor may be located on the steering wheel.

Another type steering mechanism is the caster steer. This type of wheel does not protrude outside the truck body when turning at its maximum angle.

When steering wheels are set wide apart, they are not exactly parallel when making a turn. The outside wheel is not at the same turning radius as the inside wheel and this has to be taken into account (fig. 4-10).

50 | Lift Trucks

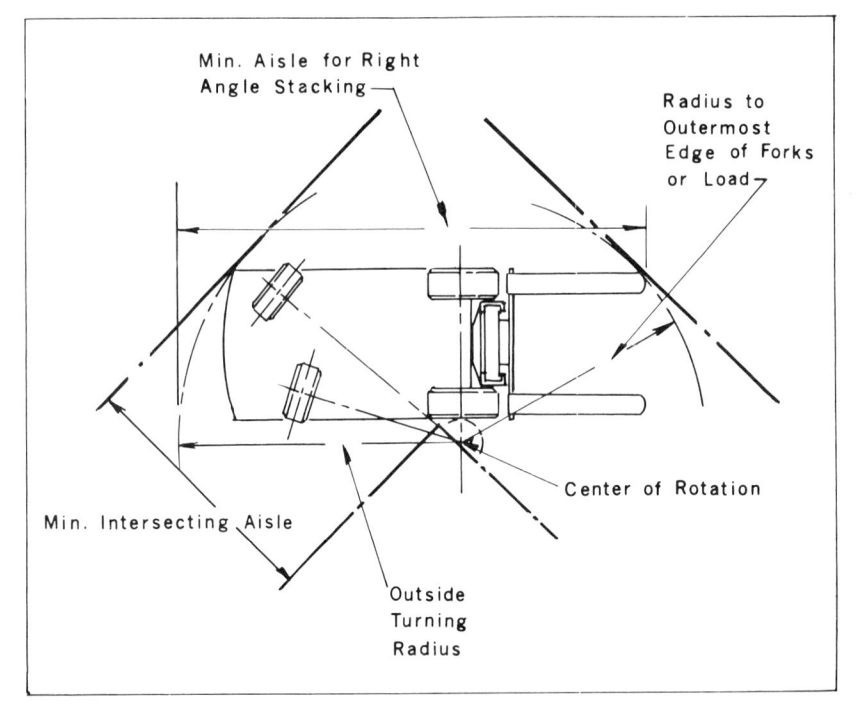

FIG. 4-10. Rear wheels are not parallel when the truck makes a turn.

Brakes

With the massive weight behind a moving lift truck, good brakes are an absolute necessity. In addition, the brakes will only be as effective as the maintenance program behind their upkeep.

Two separate braking systems are used on gas trucks, a parking brake and a brake to be used when the truck is in motion. Unfortunately, maintenance on the parking brake system can be easily overlooked. However, the parking brake is just as important as the other brakes. Often a lift truck will be parked on a ramp where the truck would gather quite a lot of momentum if it ever started rolling.

The parking brake is a shoe-type mechanical system that is manually operated by the operator. Some parking brakes are mounted on the differential input shaft instead of on the wheels.

Brakes for a truck in motion are usually the self-adjusting and energizing shoe type. A self-energizing brake is designed so that the friction

Lift Truck Design | 51

of the lining on the drum tends to pull the brake on tighter. A shoe brake is an internal brake and consists of two shoes that rub against a wheel drum (fig. 4-11). Actually, there is a lining, usually made of asbestos, which is attached to the shoes. They may also contain lead, tin, copper, and zinc.

The lining provides a "coefficient of friction" with the drum. Friction turns vehicle or machinery motion into heat, thus providing the stopping action. As a result, brake linings will wear and have to be adjusted periodically and eventually they will have to be replaced.

Most linings are fastened by brass or copper rivets. One third of the

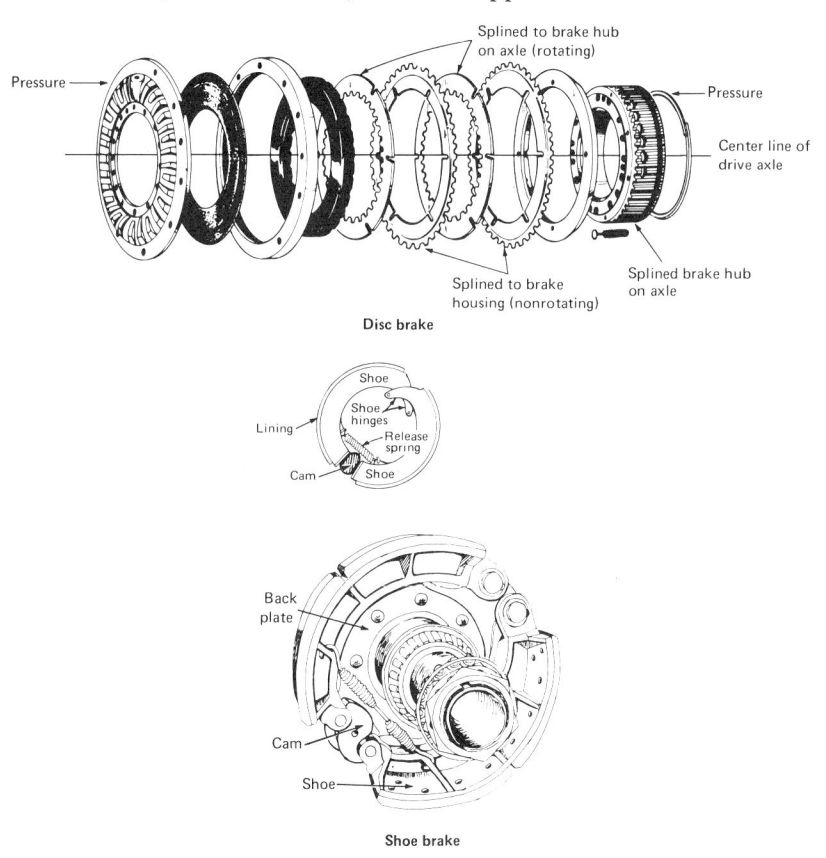

FIG. 4-11. The difference between a shoe brake and a disc brake is the method by which force is applied against the axle. A shoe brake is applying pressure perpendicular to the forces of rotation, while a disc brake is applying pressure parallel with the forces of rotation. Courtesy of North Castle Books.

band thickness is above the rivet head and the band should be replaced as soon as the lining is worn down to the rivet head. If the brake lining is not replaced, the rivet heads may cut grooves into the drum. The grooves, in turn, will ruin a new lining and the drum will either have to be machined or replaced. Therefore in the long run proper brake maintenance is the best policy.

Disc brakes, sometimes used in lift trucks, consist of a series of alternately placed lined and unlined discs. The unlined discs are splined to a stationary housing, while the lined discs rotate with the axle. Braking is obtained from the friction between the discs when they are brought together. Brakes are located on the drive wheels and are hydraulically actuated.

Electrically powered lift trucks also use what is called dynamic braking as well as regenerative braking for slowing down the motor. This type of braking is not used to stop the truck, however, only to slow it down.

Dynamic braking torque is developed if the motor is rotating below normal speed. The motor operates as a generator and the energy developed is dissipated within the motor or in a connected load. However, the torque developed by this method becomes zero at zero speed. In dynamic braking the motor is disconnected from the power supply.

Regenerative braking torque is developed if the motor rotates above the normal operating speed. The motor is always connected to the power system and the generated power is returned to the system.

Many trucks are also equipped with a "deadman's brake." If the operator gets off the seat, the truck is automatically braked and the power is cut off. This prevents the motor from running while the operator is away from the truck.

Tires

Tire selection will be dictated by whether the truck is used indoors or outdoors. For indoor usage cushion tires are used. They are lighter and less expensive than pneumatic tires, which are usually used for outside work. However, even though cushion tires allow better maneuverability than the pneumatics, higher speeds can be obtained from the pneumatics.

Lift Truck Design | 53

Cushion tires come with many different tread designs and are available in many different materials. Naturally, the need will determine the tire type and this item should be decided upon before selecting the truck.

Pneumatic tires are the most common for trucks weighing above 8,000 lbs (fig. 4-12). They can also be used indoors; however, they should not be used in areas where sharp objects could puncture the tires. Pneumatic tires are used where a high degree of traction is required, such as on ramps or in towing operations. Often dual pneumatic tires are used on each side of the front axle for traveling over soft ground, wooden flooring, or other surfaces where the available ground pressure is low. When duals travel from the outside to inside, one disadvantage is their tendency to carry stones and other debris from the yard into the building.

It is not possible to convert a truck from cushion tires to pneumatics, although the reverse can sometimes apply.

Human Engineering

In the olden days no one cared much about the driver. He was given a truck and told to go to it. However, now things are different. A driver will be more efficient if the truck is designed for the operator as well as for lifting loads.

First of all, power assists, such as power steering, power brakes, and automatic transmissions allow the operator to do more work with less effort. The increase in productivity will often more than make up for any increase in truck costs.

A comfortable adjustable seat should also be furnished. It should not be so comfortable that it will put the operator to sleep, but it shouldn't be so hard that he has to get off the truck frequently.

Controls should be located within easy reach and should be directional. That is, if the mast is tilted forward, the lever used to accomplish this should also move forward. Often, different type knobs are used so the operator will know immediately which control he has his hand on without looking at the knob.

Walkie lift truck controls should be scrutinized just as closely as a rider truck's controls (fig. 4-13). Most importantly, any emergency stops or safety switches should be placed within easy reach of the operator.

54 | Lift Trucks

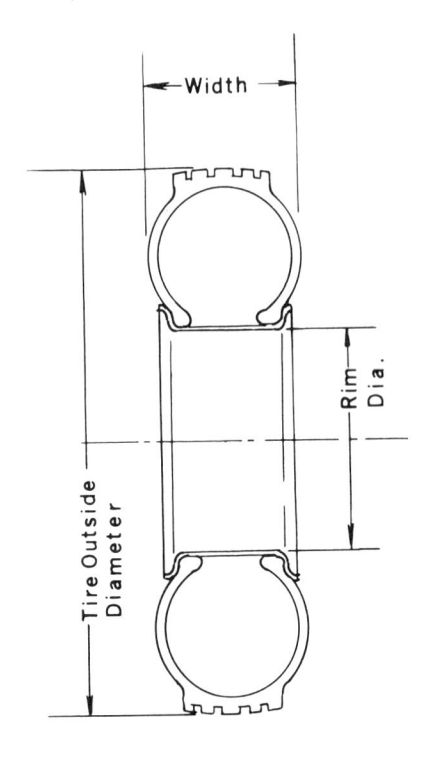

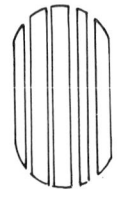

FIG. 4-12. Cross section of a tubeless pneumatic tire, the type most common on trucks weighing above 8,000 lbs. Size is usually stated as width by rim diameter: 6.70 × 15 means that tire is 6.7 in. wide and has rim diameter of 15 in. Footprint shows tire's contact with ground. By measuring area of the four tire footprints and dividing the total into truck's weight, the pressure exerted by truck (psi rating) can be obtained.

Controls should also be spring loaded to return to neutral when released.

Gages should be conveniently located where they can be seen and easily read. Some companies have gone one step further by molding symbolic identification markings on all gages (fig. 4-14). This allows the operator to know at a glance just what is happening to his truck.

Color of the truck should not be offensive to the driver (pink, for instance), and should not blend in with its surroundings. If it

Lift Truck Design | 55

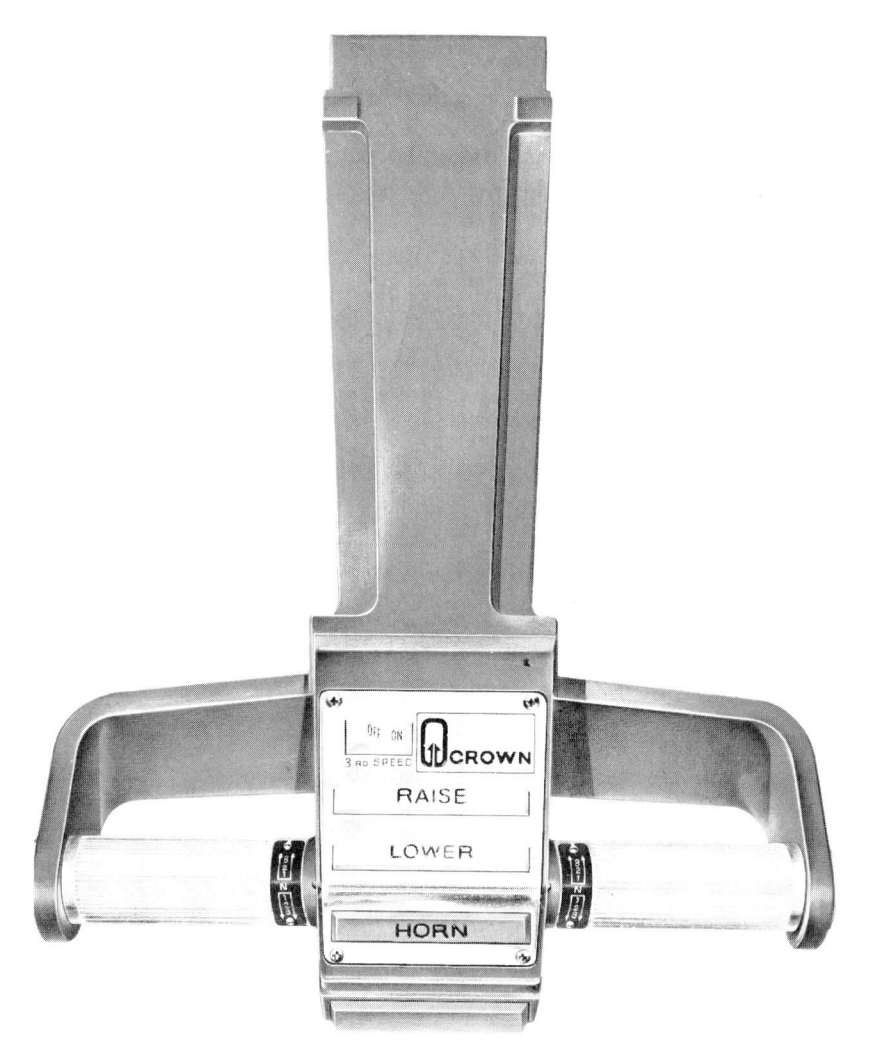

Fig. 4-13. A control handle of a walkie-type lift truck should have a safety stop botton. Courtesy of Crown Material Handling Division, Crown Controls Corporation.

does, some flashing lights or warning horn may be required on the truck.

Other human engineering features, such as visibility through the mast and accessibility to parts for maintenance, must be checked. Ac-

56 | Lift Trucks

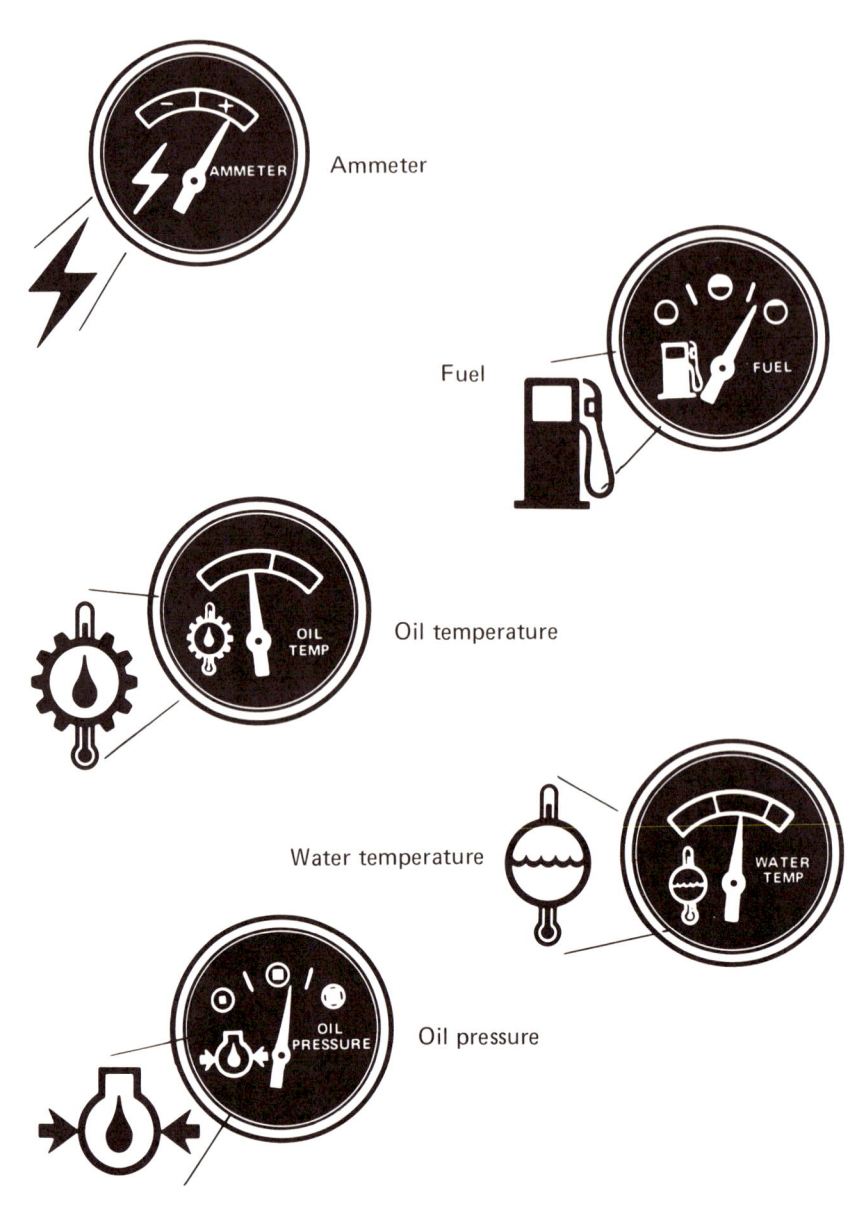

Ammeter

Fuel

Oil temperature

Water temperature

Oil pressure

Fig. 4-14. Symbolism on fuel gages allows the operator to easily identify their meaning. Courtesy of Hyster Company.

cessories may include fire extinguishers, stop lights, floodlights, and a keyed ignition switch.

If the ignition switch is keyed, be sure that the operator does not walk off with the key, leaving the second shift helpless. Usually, master keys will be controlled by a plant supervisor in a convenient location.

Most trucks are now equipped with overhead guards which are tested by the manufacturer. Therefore, it is best to obtain guards from him instead of trying to make your own. The operator's area should be kept clean and uncluttered and should offer easy-on and easy-off accessibility. This is especially important if the operator must get off the vehicle frequently, as any hindrance in his operation only increases costs.

For more detailed information on lift truck standards, the author recommends that the reader write:

American National Standards Institute, 1430 Broadway, New York, N.Y. 10018, for *Safety standards for powered industrial trucks,* B56, 1–1969, $4.50.

National Fire Protection Association, 60 Batterymarch St., Boston, Mass., 02110, for *Standard for type designation, areas of use, maintenance and operation of powered industrial trucks,* NFPA 505–1969, $1.00.

5 | Power Trains

Perhaps the hardest decision to make when selecting an industrial truck is which type of power train to use. Often the decision boils down to whether to use gas or electric power plants. There are many economical considerations to weigh (refer to chapter 6); however, in this chapter we will be mainly concerned with the operational and mechanical features of the truck. Included will be the transmission system and the industrial storage battery.

Electric trucks are noted for being quiet and fume free. Gas trucks are noted for their power and speed. Electrics are easy to operate. However, with the various transmissions now available on gas trucks, they too, are easy to operate. Gas trucks can be driven many miles before needing refueling. Yet, with the recent advances in electronics, electric trucks can also be used for many hours without needing their batteries recharged. Maintenance requirements of the two types of trucks are also quite different.

With such apparent disparities in mechanical makeup it would seem unlikely that gas trucks would compete with the electrics. Yet, the fact that they do means that the engineer selecting a lift truck should have a thorough understanding of their similarities and differences.

Gas Trucks

The power train of a gas truck consists chiefly of the engine, a clutch, transmission, differential, and drive axle (fig. 5-1). Engines may be gas or diesel, clutches wet or dry, and transmissions manual or automatic. Since almost any combination of engine, clutch, and trans-

Fig. 5-1. Power train for a gas truck is a compact package consisting of the engine, clutch, transmission, differential, and drive axle. Courtesy of Eaton Corp., Industrial Truck Div.

mission can be teamed up, power train selection can become quite complex.

Regardless of your power train selection, the following points are emphasized:

1. Select the power train that is most advantageous to the operation. For example, if driver fatigue could be a problem, select an automatic transmission. Automatic transmissions also work best where inching is required, steep ramps have to be travelled, and in towing operations.
2. Select an engine with sufficient power. If an engine with too low horsepower is selected, downtime will be higher, due to more fre-

quent overhauls, and there will be less power at all speeds. Overpower your truck—do not underpower it.
3. Select a reliable engine that can be easily maintained. Also check to see if the dealer has an adequate supply of spare parts on hand as downtime will become high if parts must be ordered from out-of-town sources.

Since most lift trucks use a gas- or LP-powered engine, this will be our starting point. A gasoline truck will have an engine similar to a car engine and should be quite familiar to most operators.

A simple review of engines shows that the driving force is created through an air and gasoline explosion that takes place in the cylinder of the engine. The force acts against the piston, which is connected to a connecting rod and crankshaft. As the piston travels up and down, its vertical motion is transferred to rotary motion through the connecting rod and crankshaft.

The piston itself makes four up and down strokes for every power stroke. The first stroke of the piston travels downward and draws an air and fuel mixture into the cylinder. The next stroke upward compresses the fuel and air mixture into a very explosive mixture. As this stroke reaches the top, a spark is introduced through the spark plug, causing the mixture to explode. The rapid expansion of gases causes the piston to travel downward, giving the engine its power stroke. As the piston reverses and heads upward again, it pushes the spent gas (exhaust) out of the engine. When the piston reaches the top of this stroke and starts downward again, another air and gas mixture is drawn into the cylinder. We call this a four-cycle engine and the strokes are as follows: intake, compression, power, and exhaust. In a four-cycle engine, the crankshaft makes two complete revolutions for every power stroke.

Two valves are located at the top of the cylinder. One valve allows the fuel and air mixture to enter the cylinder and the other valve permits the exhaust air to escape. These valves are connected mechanically to a cam shaft which raises or lowers the valves, exactly as they are, either to admit the fuel mixture or to exhaust the spent gases.

The carburetor is used to mix the air and gas so that the resulting mixture will be a proper ratio for the best possible explosion. The fuel is compressed so it will deliver a higher proportion of the explosive

Power Trains | 61

energy to the piston. Compression should be limited — the air becomes heated when compressed and, if compressed too much, becomes so hot that the mixture will explode.

A water jacket envelops the cylinder, cools the engine, and protects it from the heat generated by the combustion process.

The firing of the spark plugs is achieved through the distributor. A rotor unit located inside the distributor causes electrical charges to be sent to the spark plugs so that the plugs fire in the proper order. Most gas trucks have at least four cylinders, which are fired in a specific order to more evenly distribute the pulsations caused by the explosions in the cylinders. A flywheel, which helps achieve a smooth-running engine, is connected to the crankshaft and through its inertia keeps the crankshaft revolving.

Exhaust gas travels through the exhaust valve into the exhaust manifold, and next into a muffler, and then is forced into the air. The muffler contains a series of baffles which help to quiet the noise of the engine.

Mufflers are made in various types, depending upon the requirements of the particular job. Most exhaust gases contain a certain amount of incompletely burned fuel — carbon monoxide (CO) — which becomes a very lethal gas if it is allowed to concentrate to any extent in normal air. As a result, some mufflers contain a catalytic material (fig. 5-2), which helps to change the CO to nonpoisonous

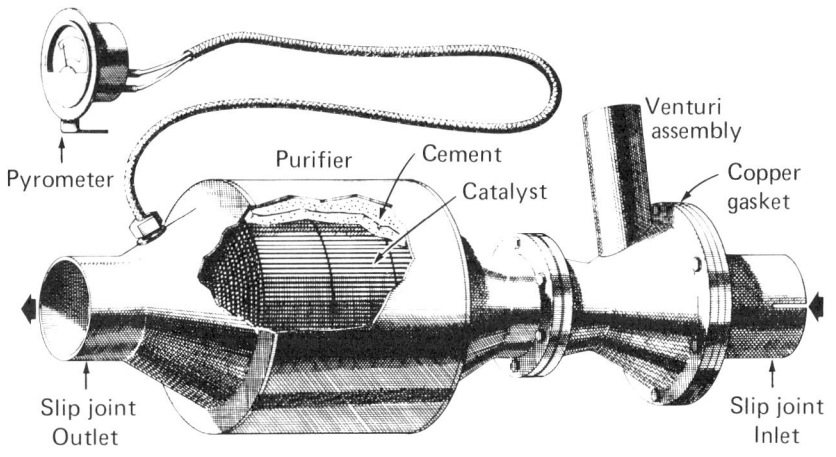

FIG. 5-2. A catalytic muffler can help reduce engine pollutants. Courtesy of Oxy-Catalyst

62 | Lift Trucks

carbon dioxide (CO_2). Another type of muffler contains water, which quenches hot particles, reduces exhaust temperatures, and virtually eliminates sparks.

In recent years, many users have converted their lift truck operations to liquefied petroleum gas (LP gas) operations. The main constituent of LP gas is propane, which is an extremely clean burning fuel. Kits are available for use in converting most trucks to LP gas and new trucks can be obtained ready for LP gas usage (fig. 5-3).

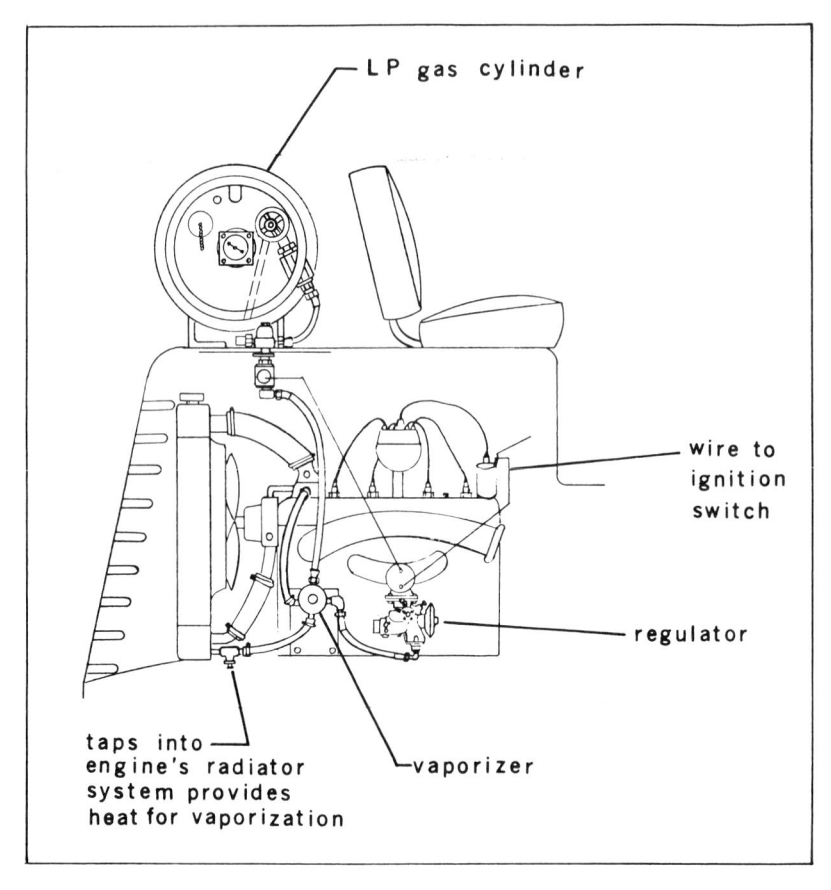

Fig. 5-3. Many lift trucks can have an LP gas conversion unit installed at the factory. Taps from vaporizer into the engine's radiator system provide heat for vaporization. Courtesy of J & S Carburetor Co.

The concentration of CO in the exhaust of a propane truck is very low when contrasted to a gas fueled truck and, as a result, a propane truck can be used in almost any inside building area.

Since propane is a cleaner burning fuel than gasoline, truck downtime is usually less. A safety feature of the propane fuel truck is that it has a closed type of fuel system. If the fuel line should break, special valves in the line automatically shut off the fuel, preventing it from leaving the tank. Propane has a higher thermal efficiency than gas and, as a result, its use may reduce the overall fuel costs.

However, propane fueled trucks are not without their hazards. Propane is a combustible gas at high pressure and, as such, should be treated with respect. Chief hazard is the opening of the relief valve on the fuel container.

Other hazards stem from operating the truck in improper areas and from poor maintenance. Propane fuel trucks are not recommended for high temperature areas, such as near ovens and furnaces, except for extremely short periods of time. It goes without saying that propane trucks should not be stored or left overnight in these areas.

Poor maintenance could lead to carburetor leakage. There is also the possibility of leakage during refueling. A quick-disconnect coupling system should be used in the fuel line between the tank and truck.

Procedures for proper storage and handling of LP gases may be obtained from the National Fire Protection Association or from the Engineering Division of the Associated Factory Mutual Fire Insurance Companies Association.

Diesel engines are quite similar in operation to gasoline engines except that they do not have spark plugs. As mentioned previously, a gas tends to heat up when compressed. However, in a diesel engine the fuel isn't mixed with the air through a carburetor. The compression ratio (which is the ratio of the air space in the cylinder at the bottom of a stroke to that at the top of a stroke) is so high in a diesel that the air temperature may be 1,000° F. As the piston compresses the air in a diesel engine, the fuel is injected in exactly the proper amount at just the right time so that the hot compressed air ignites the fuel, causing the power explosion. Because of the higher temperatures and compression, a diesel engine must be stronger and more accurately built than a gas engine. In addition, the metering system for introducing the fuel into the cylinder must be very exact. Often in small models, diesel engines cost about twice as much as a comparable gas engine. However, this price difference diminishes as the size of the engine increases.

Diesels are usually more economical than gas engines as the fuel costs less and, with the higher compression in the engine, the fuel has a higher heat value, therefore more efficiency. A diesel is normally a clean burning engine and its exhaust is almost completely free from carbon monoxide. There are bad smelling and slightly toxic gases in the exhaust which may be objectionable to some workers. However, a diesel exhaust should not be dirty. A smoky exhaust will indicate too rich a fuel mixture or else defective parts.

Without an ignition system or carburetor, maintenance is somewhat less than that for a gas engine. Another advantage of the diesel is that, with the use of a nonvolatile fuel, a diesel-powered vehicle can be used safely in hazardous areas such as cotton warehouses, and chemical, paint, and petroleum plants.

Diesel engines ordinarily are not converted to LP gas. The conversion is quite expensive and there is usually no need, as diesel fuel itself is less expensive than gasoline.

A diesel engine, like the gas engine, is available in a four-stroke model. However, both diesel and gas engines may also be obtained in two-stroke models. Two-stroke engines have one power stroke for every revolution of the crankshaft. In two-stroke engines, at the bottom of the power stroke ports open on opposite sides of the engine, and air (or, in gas engines, air/fuel) is admitted and exhaust gas is forced out. The compression stroke then takes place and the air (or air/fuel, in gas engines) is compressed.

The makeup of a diesel engine varies — in addition to being available with either two- or four-stroke engines, an air or water cooling system is available. Air-cooled engines are constructed with fins which are attached to the external portions of the cylinder to provide a better heat transfer surface and to direct the air flow around the cylinder.

In almost all cases, a gasoline-powered lift truck will be equipped with a four-stroke, water-cooled engine.

Although this review has made the gas and diesel engines seem rather simple, remember there are many components to any engine and they must all be thoroughly maintained for maximum efficiency.

The power from the engine has to be transmitted to useful power at the wheels of the truck, which is the function of the other parts of the power train mentioned previously. In the simplest terms, power from the engine goes to a transmission which converts high speed engine rotation into a lower speed, useful, working rotation. The power goes from the transmission into the differential, where it is transmitted to the wheels.

A clutch, located between the engine and transmission, provides the means of engaging and disengaging the power between the two units. Clutches may be either a wet or a dry type. Basically, the clutch consists of a springloaded plate with the clutch-driven disc located between it and the flywheel. The springs push the pressure plate against the disc which, in turn, squeezes the plate between it and the flywheel. The friction between the flywheel disc and pressure plate is sufficient to transmit the rotating motion of the flywheel to the pressure plate which drives the jack shaft or pilot shaft. The jack shaft then transmits the power to the transmission. By retracting the pressure plate, the power is no longer transmitted to the jack shaft.

A wet clutch or oil clutch (fig. 5-4) is used when there are severe "stop and go" operations. It differs from a dry clutch in that oil is continuously sprayed on the clutch pressure plate and flywheel to dissipate the heat that rapidly builds up in a clutch that is used heavily. A wet clutch provides better "inching" control and eliminates chatter and vibrations.

Power is transmitted from the clutch to the transmission. The transmission is a set of gears and shafts which provide a change in the speed-power ratio (fig. 5-5). Most transmissions have two or more different forward speeds and two reverse speeds. Reverse action is obtained by mounting a shaft called the reverse, or idler, shaft parallel to the main shaft and the counter shaft (the shaft mounted parallel to the main shaft that has the various speed gears on it). The gear on the idler shaft meshes with a reverse sliding gear on the main shaft and a gear on the counter shaft to provide reverse motion.

There are various gear trains built into lift trucks, including bevel gears, planetary gears, worm gears, etc. Chains and belts can also be used to transmit power. Rather than go into explanations of them here, the author recommends reading separate material on mechanical power transmission. The main point to remember is that gears are mounted on shafts which, in turn, rotate within their supports. Since there is motion involved, friction must be reduced. If your transmis-

66 | Lift Trucks

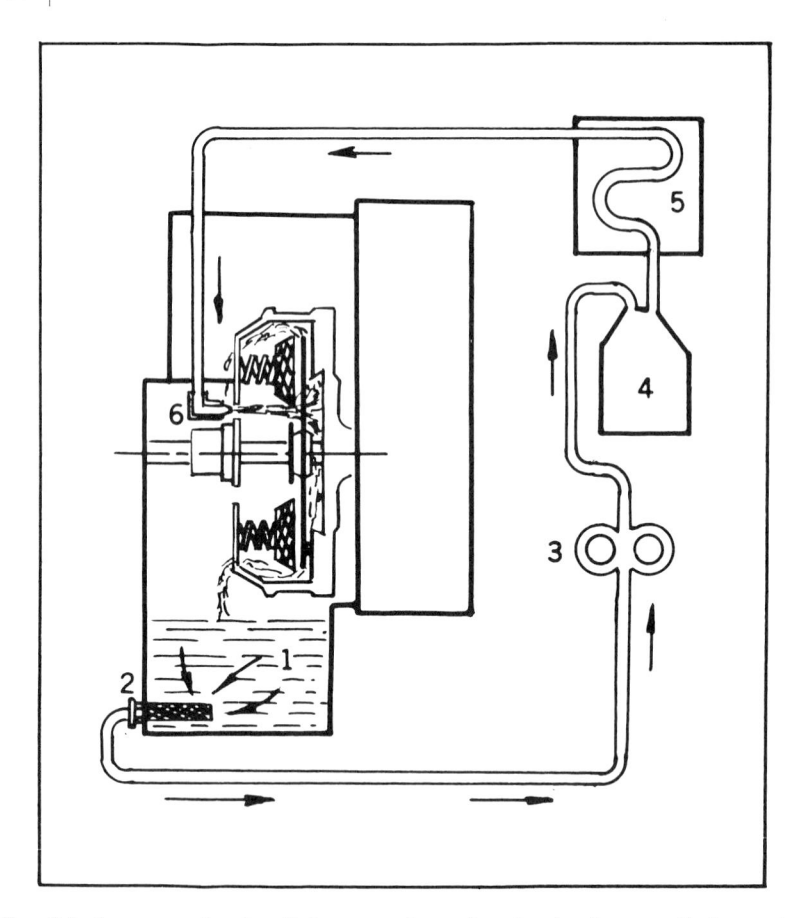

FIG. 5-4. In a wet clutch, oil is sprayed on the clutch plate to dissipate the heat build-up that occurs when a clutch is used heavily. The sump (1) provides for storing and cooling the oil. Oil entering the sump passes over a deaeration baffle (2) and then through a screen (3). Either a vane or gear pump (4) pumps the oil through a filter (5). The filtered oil passes through a cooler (6) before going through the nozzle (7). The oil is sprayed across the pressure plate of the clutch.

sion and bearings are not permanently lubricated, be sure that scheduled lubrication is carried out on all these components. Remember, dirt is always the enemy of any moving surface.

In order to reduce driver fatigue, many lift trucks are now equipped with automatic transmissions. Through the use of hydraulic torque converters, power is transmitted by a fluid rather than through me-

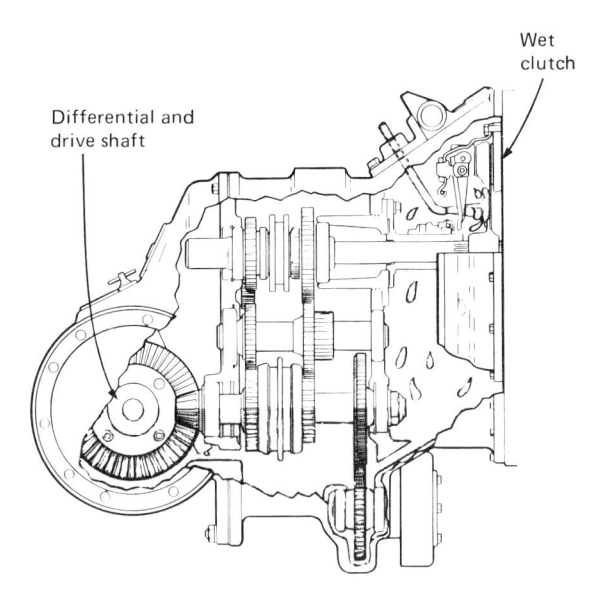

FIG. 5-5. A standard transmission is illustrated with a wet clutch. Courtesy of Clark Equipment Co., Industrial Truck Div.

chanical gearing (fig. 5-6). The torque converter is a liquid turbine consisting mainly of an impeller that is attached to the flywheel, a freewheeling stator, and the turbine, which is splined to the output shaft. The converter is completely filled with oil and is kept under constant pressure by a hydraulic pump that is driven by a gear in the transmission portion of the system.

In operation, the impeller (or really the pump), spinning with the flywheel, picks up oil near the center and hurls it against the turbine vanes at the outer edges. This motion causes the turbine to rotate, which in turn causes the shaft to rotate. The stator provides curved passages so that oil leaving the turbine will change direction and re-enter the impeller (the pump). This regenerative action is the key to the torque multiplication developed in the converter. Thus the velocity of the oil leaving the stator is added to the velocity of the oil developed in the impeller so that the velocity of the oil leaving the impeller is correspondingly greater.

The main advantage of automatic transmissions is the elimination of

68 | Lift Trucks

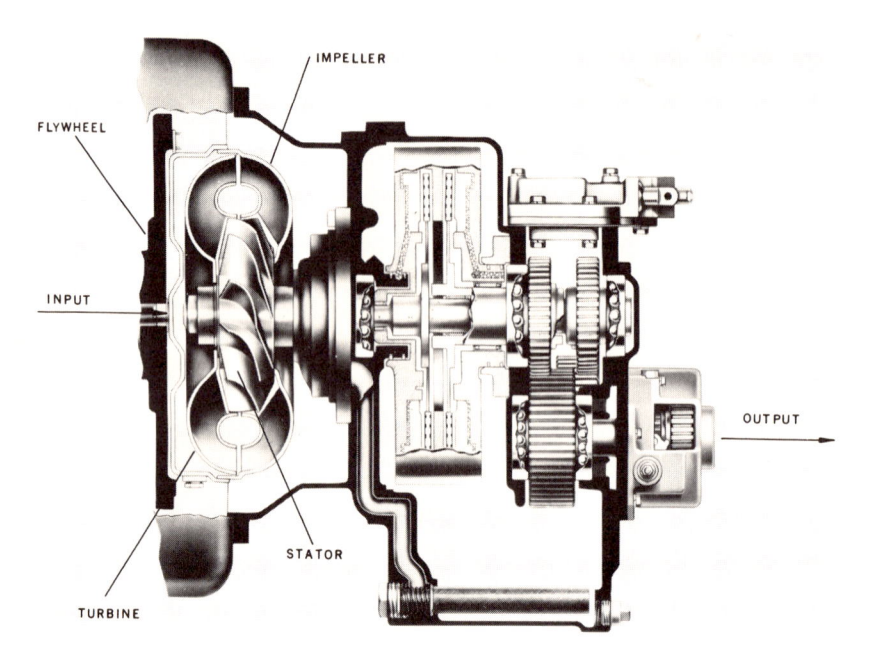

Fig. 5-6. Power in an automatic transmission is often transmitted by means of a liquid turbine. Courtesy of Allis-Chalmers.

the clutch. "Inching" is accomplished more easily with automatic transmissions and, even though it may cost more, the trade-in value will be higher. Many variations of automatic transmissions are available, for example, the hydrostatic transmission shown in figure 5-7. It is a self-contained unit without a clutch or gears, yet it can transmit high engine flywheel inertia to the drive wheels.

Power has to be transmitted from the transmission to the wheels. This transmission action takes place in the differential. A bevel gear carries rotation around a right angle and can also change the speed power ratio at the same time. The bevel gear assembly in a differential is quite complex. The main gear, the ring gear, turns the axle and, in turn, is driven by a pinion gear which is attached to the shaft from the transmission. However, since there are two wheels being driven on the drive axle there is another set of gears inside the ring gear. This set of gears compensates for the different speeds the wheels revolve when going around corners. Actually, when going around a corner the outermost wheel travels farther than the inner wheel. This inner set of gears

Power Trains | 69

Fig. 5-7. An automatic transmission without a clutch of gears is this hydrostatic transmission. Courtesy of Towmotor

consists of six bevel gears with two of the gears attached to the inner ends of the axles. The other four gears are called spiders and are mounted in opposed pairs by a spider shaft assembly (an X-shaped shaft) attached to the ring gear (fig. 5-8).

The spider gears rotate with the ring gear. They rotate the axle gears causing the whole system to revolve on the axis of the axle (fig. 5-9). If one wheel should lock, then the spider gears, in addition to rotating around the main axle, will rotate around their own axis on the X shaft. The other axle continues to be driven by the rotation of the spider gear assembly as well as by the spider gears, therefore, its speed is doubled. As a result, if one axle is slowed, the other axle is speeded up proportionately.

One fault of a spider gear assembly is that the wheel that offers the least resistance is the one to which the most of the power is applied. As a result, if the wheel becomes locked on dry land, and the wheel that is spinning is on ice, then the spinning wheel will receive all the driving

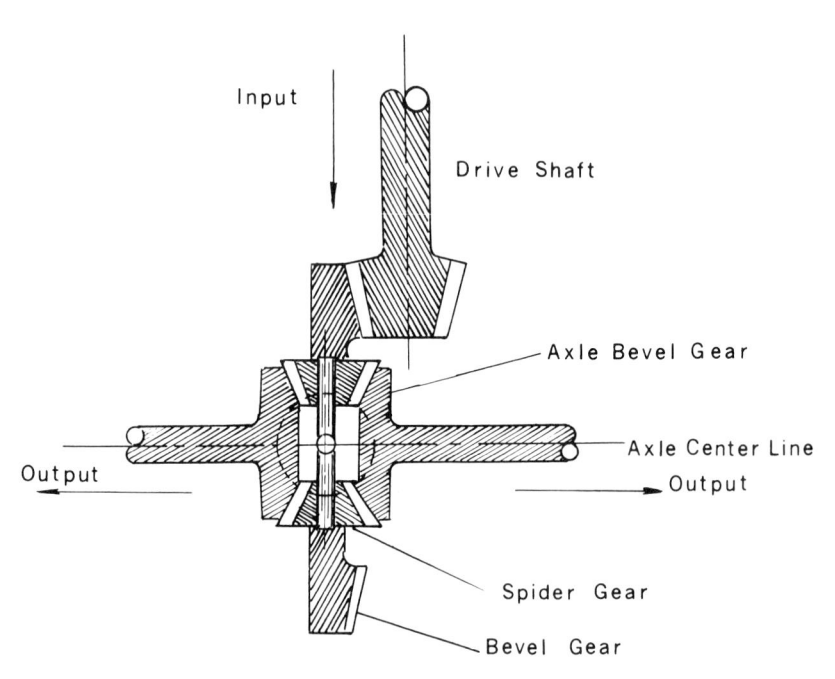

FIG. 5-8. In the differential, bevel gears are used to transmit power from the drive shaft to the two drive wheels.

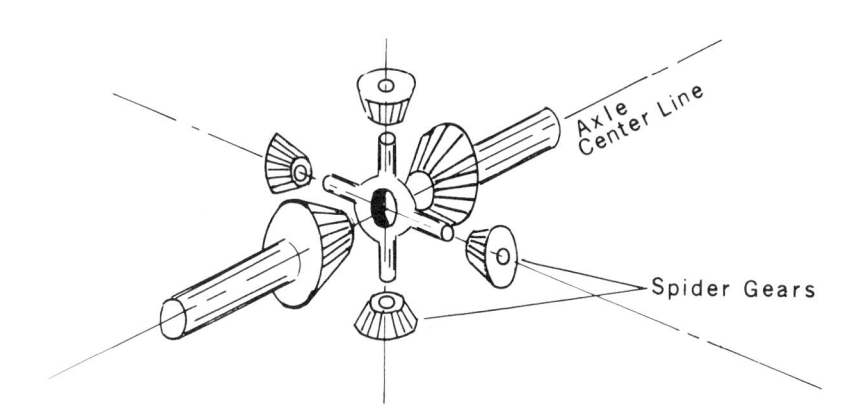

FIG. 5-9. The spider gear assembly in the differential allows one wheel to travel faster than the other when turning a corner. Without such an assembly, one wheel would tend to skid when turning.

force. This situation would hold true in loose sand or dirt and would cause a truck to be unable to move if it were stuck. As a result, one may obtain a no-spin differential as an added feature on his truck. This type of unit features a clutch unit on each side of the spider assembly instead of spider and bevel gears.

In addition to all the gears in the transmission and differential, there are also gear assemblies in the wheel drums. Usually, they consist of a spur gear arrangement or planetary drive. The spur gear system allows for more clearance under the axle housing; however, the planetary drive gives greater tractive effort at the wheels and is also more durable (fig. 5-10).

Usually on the smaller lift trucks, the power train is assembled in one oil-tight housing. However, in the larger trucks, the engine, clutch, and transmission are in one assembly. This assembly is located at the back of the truck with the power being transferred to the differential in the front by a drive axle and universal joints, as in a car. The latter method offers a better weight distribution in the truck.

Electric Trucks

For a long time, electric power wasn't even considered in many industrial truck applications. Batteries for the trucks were expensive and

72 | Lift Trucks

Planetary gear train

Spider gear assembly in differential

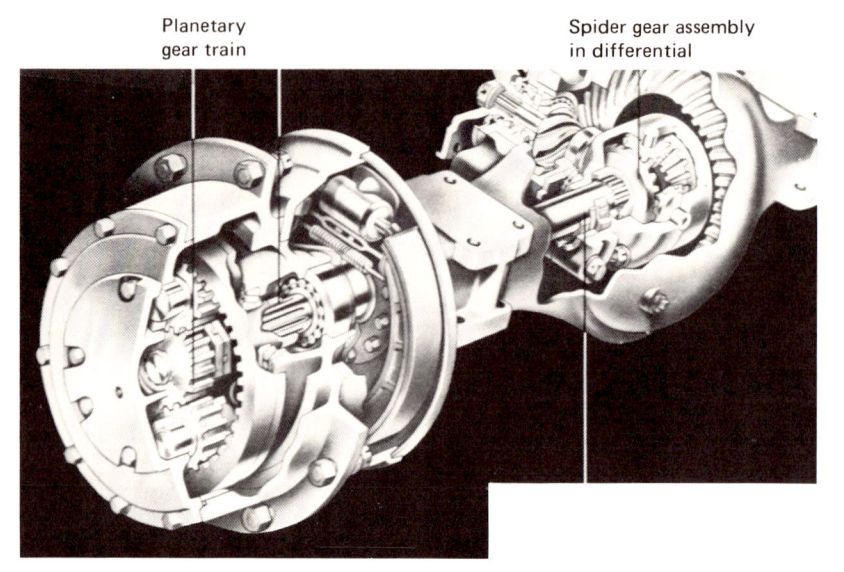

FIG. 5-10. A planetary gear system at the drive wheel allows greater traction than a spur gear system. Courtesy of Clark Equipment Co., Industrial Truck Div.

had to be frequently recharged. Electric controls consumed a great deal of power and motors were not powerful enough for most applications. However, improvements in controls and in items like battery leasing plans plus many other truck improvements have now made electric trucks quite competitive with gas trucks.

The following are probably the four most important items to consider when selecting an electric truck: motors, controls, batteries, and battery chargers.

Even though the dc motor is the heart of the electric truck, it is the item that is probably of least concern to the user. It is very compact, requires little maintenance, and fits directly on the truck's drive axle (fig. 5-11).

Motor design is one of the most important improvements in the electric truck industry. Previously it was thought that electric lift truck motors should be totally enclosed and nonventilated. The reasoning was that an open protected motor would become easily contaminated because of its location in the bottom of the truck.

However, this is not the case and open protected motors are being used in a number of applications. A modification of this motor is where the top half is covered by solid metal and the bottom half of the motor is covered by an expanded screen.

Nevertheless, a totally enclosed explosion-proof motor is required in areas containing explosive gases and volatile liquids. Such motors should meet Underwriters' Laboratories specifications for Class I, Group D or Class II, Group C ratings.

Electric motors are chosen by their torque rating at a particular voltage and amperage. Horsepower rating is not too important in electric motors, as a horsepower rating may be designed for one of several voltages depending upon what battery or what other motors are used in the system. For example, the horsepower rating for an electric-powered truck may be only one ninth of the horsepower rating of a gas truck with similar capacity.

Most electric truck drive motors are series wound because of their high starting torque capability. A straight series motor will "run away" under light or no-load conditions; however, this isn't a problem in electric trucks as the propulsion motor is permanently connected to a load through the gear system and cannot really be unloaded.

Almost all industrial truck motors under 3 hp operate on a 12-volt battery. Motors over 3 hp run on a 30- or 36-volt system, although higher voltages are in use — maximum battery rating is about 72 volts.

Although the propulsion motor may have only a rating of from $\frac{1}{4}$ hp to 15 hp, the pump drive motor may have a rating from 3 hp to 20 hp. This would seem rather unusual, but there is a simple explanation. A motor is also rated on the length of time it runs. In order to meet larger capacities in lifts as well as the high speeds for making those lifts, a high horsepower motor is required. But, since the motor runs for only a relatively short length of time, it can actually be physically smaller in size than the drive motor. The exception to this case would be a power steering motor as it has to have continuous rating.

As mentioned previously, series wound motors have a high starting torque. This is because the torque is proportional to the square of the current, resulting in the torque being its highest when speed is the slowest.

Two other dc motors should be mentioned: the shunt and the compound motor. In the shunt motor the torque is proportional to the current and, as a result, is practically a constant speed motor. The compound motor is a combination of the series and shunt and has a lower

74 | Lift Trucks

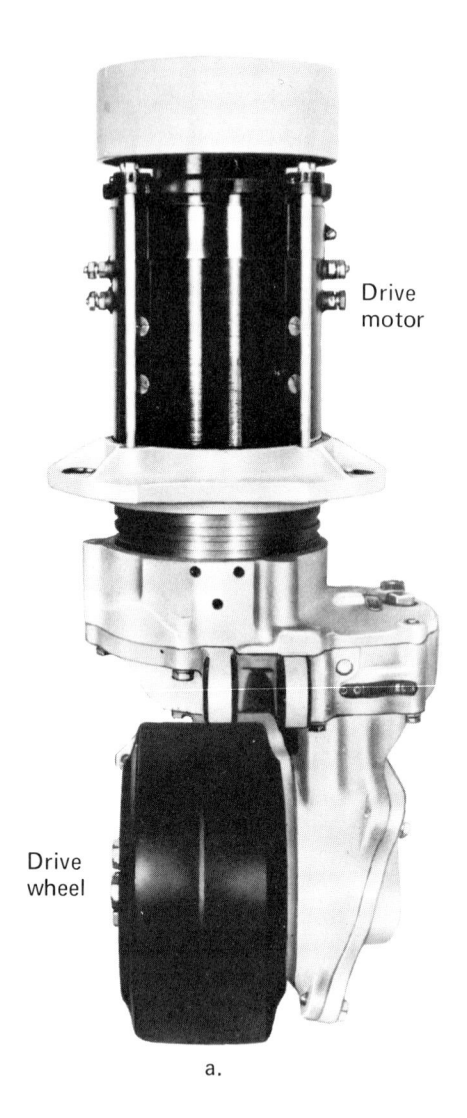

FIG. 5-11. Electric motor location in a walkie truck (*a*) and in a standard lift truck (*b*). Courtesy of Barrett Co., and Otis Material Handling, Otis Elevator Company.

Power Trains | 75

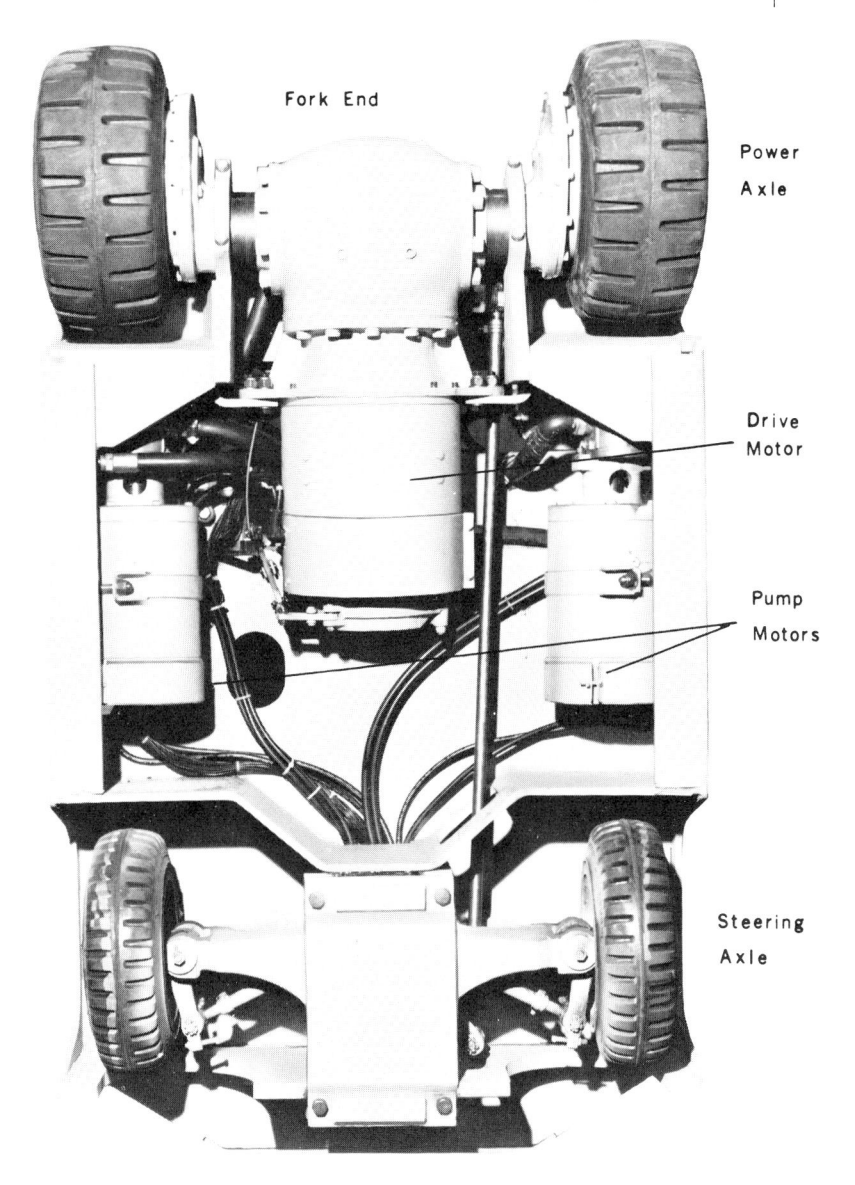

BOTTOM VIEW — ELECTRIC LIFT TRUCK
b.

starting torque along with less variation in speed when the load is increased. Because of this, compound motors are sometimes used to drive the hydraulic pumps since they do not develop excess motor speeds under no-load conditions.

As mentioned in chapter 4, one advantage of the electric motor is that it can be of great assistance when braking. When an electric motor is driven, it acts as a generator and produces a current, which, in turn, brings about a braking action on a loaded motor.

If you need a truck that is capable of making frequent starts and stops, then you will need a truck that has one of the newer power-saving control devices.

The speed of a dc-series motor is controlled by varying the voltage across the terminals of the motor. Since a constant voltage battery is used, then some external element is used to vary the motor voltage. The simplest control is a switch with which to connect and disconnect the battery to the motor load. However, this is not practical since little speed control is provided.

Most lift trucks have been and many still are equipped with a stepped resistance control. The resistors provide four accelerating steps and at full speed the battery is connected directly to the motor. A refinement of the step control is the carbon pile. This type of control consists of a series of carbon discs placed together to create resistance. Pressure applied to the discs squeezes them together, thus varying the resistance value and changing the speed of the motor.

Both the carbon pile and stepped resistant control are dissipative types. As such, the resistant control draws full battery power at all speeds and dissipates unneeded power at low speeds through resistant devices. Therefore, at low speeds this type of control is quite inefficient.

An alternate method for controlling speed is by rapidly turning the current on and off. Through the use of solid state devices, the current can be switched on and off with very little waste of power (fig. 5-12).

Solid state devices control the speed by pulsating the current to the motor (fig. 5-13). Pulses are varied in frequency, with each pulse momentarily energizing the motor. The motor acts as a flywheel when the current is off and its momentum keeps it rotating until the next pulse. At about 70 percent of full motor speed, the solid state unit cuts out and the current flows directly from the battery to the motor.

Therefore, the more a truck operates at low speeds, the more efficient

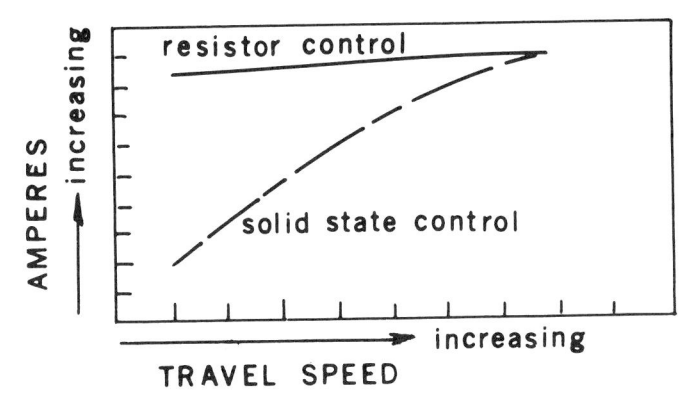

FIG. 5-12. Solid state control is desirable for narrow aisle truck applications since their runs are short and frequent stops and starts are required. Chart compares power draw between solid state and resistance controls. Solid state provides smoother inching and acceleration plus more operating time between charges.

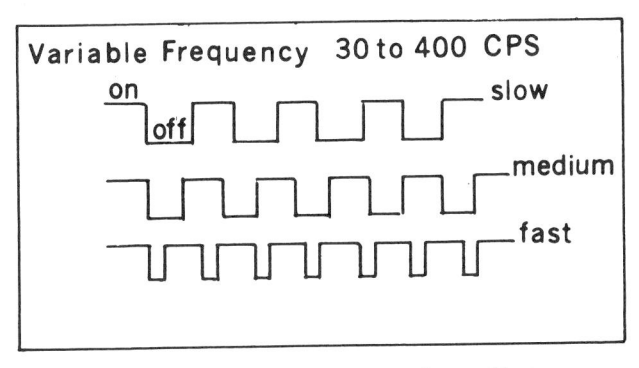

FIG. 5-13. A solid state unit requires less current than a resistor type of control as speed is regulated by rapidly switching the power on and off. The solid state control is desirable as it requires less draw on battery current. Illustration shows how system is switched on and off for controlling speed.

a solid state control is over a resistant type control. Energy losses within the solid state system are less than 5 percent at all speeds. As a result, the solid state device will allow a longer operating period between recharges.

Just as the gas engine needs a fuel, the electric motor also requires its "fuel." In this case we are talking about the large storage batteries mounted on the electric truck.

Storage batteries must be charged before they can generate electricity. The battery does not store electrical energy, rather it converts the charge to chemical energy. The two most common batteries used in lift trucks are the lead-acid and nickel-iron-alkaline battery. They both operate on the same basic principle; however, they do differ in materials and characteristics.

A storage battery is made up of many individual cells that can be arranged through different connections to provide various voltages and current capacities. Basically, each cell has a potential of 2 volts. Therefore, if a 24-volt battery were desired, it would contain 12 cells. The ampere capacity of each cell is determined by the surface area of the plates in the cells.

How does the storage battery work? Using the lead-acid battery as an example, electrodes or plates of different materials are immersed in a solution which is called the electrolyte. Any time two dissimilar metals are immersed in an electrolyte and a circuit is closed between the two, a flow of electrons is created. In the lead-acid battery, lead peroxide is used as a positive plate and sponge lead for a negative (fig. 5-14).

The lead-acid battery electrolyte is made up of sulphuric acid and, as the battery is discharged, lead sulphate is deposited on the plates. The acid loses its strength and its specific gravity approaches the specific gravity of water. As a result, the strength of a battery can be easily determined by using a hydrometer (a device that measures specific gravity).

Battery capacity is expressed in ampere-hours. This capacity is determined by taking the number of amperes required and multiplying them by the actual number of hours the battery can be used.

The nickle-iron battery has a positive plate made of various nickel materials and a negative plate made of iron oxide; the alkaline electrolyte is noncorrosive. A nickel-iron battery is more durable than a lead-acid battery; it also costs more.

To offset the cost, the nickel-iron battery has many features not found in lead-acid batteries. The electrolyte is noncorrosive. If the nickel-iron battery is accidentally discharged, there is no permanent damage; merely recharge it and it is ready to go. Another advantage is that the nickel-iron can be stored indefinitely without deterioration.

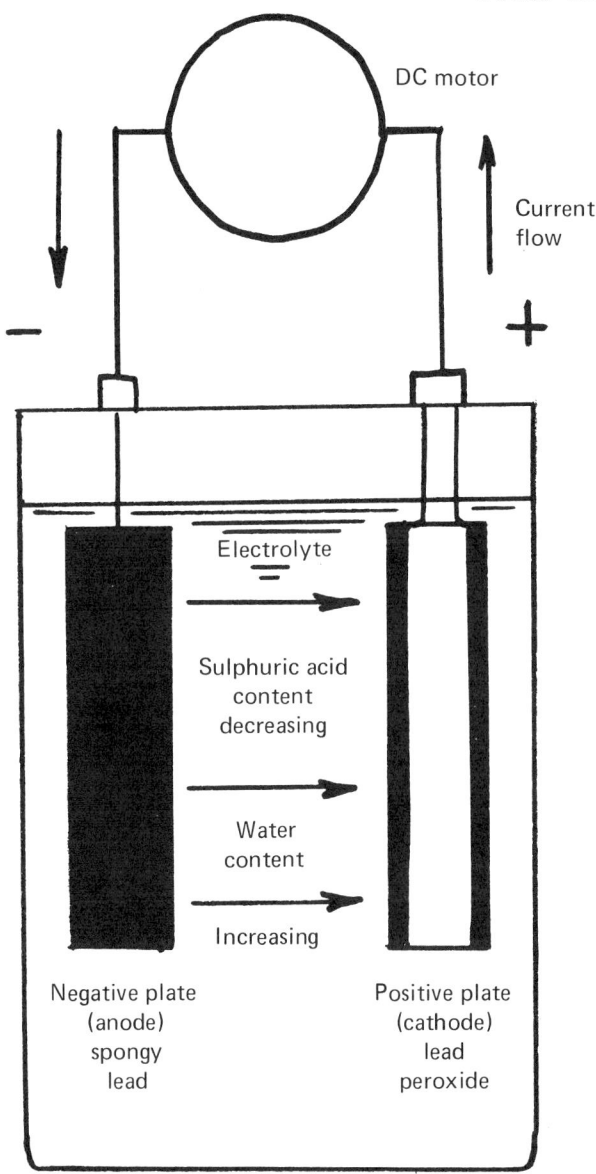

Fig. 5-14. Current from a charged storage battery is the result of a chemical action and not the "storing" of an electrical charge. Upon discharge, the chemical action of a storage battery involves the cathode, which is PbO_2; the anode, Pb; and the electrolyte, H_2SO_4. At the cathode or plus terminal, the following chemical action takes place: $PbO_2 + H_2SO_4 = PbSO_4 + O_2$. At the anode, or minus terminal, the chemical action is: $PbO_2 + H_2SO_4 = PbSO_4 + 2H_2 + H_2O$.

80 | Lift Trucks

However, lead-acid batteries have been extremely popular for many years. In addition to industrial uses, they are used in all car batteries. The lead-acid battery has more power potential per pound than any other battery and costs less. Recent developments include longer life and the use of epoxy for cell-jar sealing (instead of an asphalt compound).

Among the more recent battery developments is the fuel cell. Although not presently used in lift trucks, its output possibilities are much greater than those of the electrolyte type of battery. The fuel cell generates electricity through the chemical reaction of common fuels. As a result, the fuel cell does not need recharging, but it does require the fuel to be replenished. It is expected that once the fuel cell becomes competitive in price, whole new areas will be opened up to dc motors in fields now dominated by the gas engine.

The following information should be gathered to help determine the battery size, as these data affect current requirements: the weight of the truck; average weight of loads handled; the average length of each trip; number of trips per hour; number of lifts per hour; number of hours the truck is used per shift and any unusual requirements such as going up ramps or frequent accelerations. As an example, if the truck uses 200 watt-hours per trip and makes 100 trips per day, then it requires 20,000 watt-hours or, dividing by 1,000 to get kilowatt-hours, the truck consumes 20.0 KWH per day. Usually, a 25 percent safety factor is used, resulting in a requirement of a 25-KWH minimum battery size.

When selecting an electric lift truck remember also to select a battery charger and provide a convenient area in which to use it. The area should be kept clean and well ventilated. Sufficient space should be provided for the trucks, spare batteries, and for the chargers. Often, some piece of handling equipment, such as an overhead hoist, will be required to move the batteries.

Because a battery is unidirectional in current, dc current is required for recharging the battery. In order to use ac current for recharging, the ac current is used either to drive a dc generator or the current must be rectified to dc current.

The following methods are used for recharging:

1. Modified constant potential — the direct current voltage is held at the rated voltage of the battery. The charging current is automatically

reduced as the charge progresses. When the charge is complete the equipment shuts off. The formula for calculating the kilowatt requirements for a motor generator to charge batteries in multiple from a fixed voltage bus in 8 hrs is:

$$KW = \frac{A.H. \times 0.225 \times \text{no. cells} \times 2.25 \times 0.8 \times \text{no. bat.}}{1,000}$$

Both lead-acid batteries and nickel-iron batteries can be charged by this method. For general use, this method ensures uniform and dependable charging and is probably the best method.

2. Taper method — a variation of the modified constant voltage method. It can be used on either lead-acid or nickel-iron batteries; however, it is employed only where one battery is charged at a time. The formula for calculating the kilowatt requirements for a single-circuit motor generating set is:

$$KW = \frac{A.H. \times 0.225 \times \text{no. cells} \times 2.25 \times 0.8}{1,000}$$

In the modified constant voltage method, the current is passed through a fixed resistance in the early part of the charge, in series with the battery on the charge. In the taper method, no resistance is placed in series with the battery.

3. Two-rate charging — used for charging lead-acid batteries only by means of rectifiers. A high rate of charge is maintained until the gassing point of the battery is reached. At that point, the rate is reduced automatically to lower finishing rate. This occurs at about 80 percent of full charge. This method usually employs a temperature voltage control relay and a timer. As a result, the unit permits more resistance to be cut into the circuit when the gassing point of the battery is reached.

4. Constant current method — used for charging nickel-iron batteries by the use of rectifiers. It may be used for charging lead-acid batteries, but, it would take 12 to 16 hours or more to do so.

Battery charging may be controlled manually or automatically. Because of the potential equipment damage that can be done by manual control, automatic charging is recommended.

A recent trend is to mount the battery charger directly on the truck. This allows the truck to be recharged anywhere in the shop. However, be certain the maintenance man does not forget to check the battery for water.

Before placing a new battery into service, it should be given a freshening charge. This operation will usually require only 2 to 3 hours. Occasionally a boost charge or an equalizing charge should be given to the battery.

A boost charge is given when the capacity of the battery is insufficient to meet the need required. Such a charge should be used only as an emergency measure and the charge is usually given at a high rate for a short duration. There are three possible reasons for a battery needing a boost charge: too small a battery was selected; the duty cycle has been increased; or the battery is old and near the end of its useful life.

An equalizing charge is used to overcome any inequity in voltage or specific gravity as a result of continuous cycling over a long period. Such charges are given at a low rate and should be required only once or twice a month.

When charging a battery, one should always remember to connect the positive pole of the battery with the positive terminal of the charging source and the negative pole to the negative terminal. Serious damage will result if connections are made otherwise. In addition, a battery should always be recharged promptly — if a battery is left standing in a discharged condition, the sulphates will harden on the plates. Never allow a completely discharged battery to stand longer than 24 hours or when the temperature is below freezing.

6 | Financing

Lift truck manufacturers have been leaders in leasing and other financing methods for obtaining industrial equipment. This chapter covers that age old question: lease, rent, or buy?

There is the story about how the production personnel of a "tight fiscal policy" company managed to put one over on the accounting department. They bought a complete truck "piecemeal," as maintenance parts, and then had it assembled in the plant.

Such a procedure is possible in many companies and the above story may have a lot of truth in it for, without adequate maintenance records, labor and replacement parts can reach astronomical proportions. Although more than just lift truck parts were included, the use of computer control enabled a large pharmaceutical plant's maintenance stores to realize an inventory savings of over $100,000 the first year.

New equipment is easier to justify from a production point of view than from a financial point of view. Equipment for production needs is usually determined by making a plant layout and equipment analysis (see chapter 9). This type of analysis is under the control of personnel working for various manufacturing departments.

The financial analysis may be under the control of those not familiar with production needs. As a result, if a financial analysis is prepared by manufacturing personnel, problems will frequently be encountered when making an equipment justification presentation to corporate management for their approval.

Manufacturing personnel, therefore, should become familiar with the accounting and financial procedures of their respective companies. For example, a plant engineer at a large Midwestern foundry made an equipment analysis and, in the process, discovered that almost every lift truck in the plant was beyond its useful life and needed replacing. The results of his study indicated four courses of action. First, the company could sit back and do nothing. Their maintenance costs were already excessive, and not only were their trucks old, but replacement parts were not available for many of them. A further problem became apparent — because of their age many trucks were underpowered or undersized for their particular job application. The net result of the study showed that to continue with the existing truck fleet would almost certainly guarantee excessive downtime and production problems.

A second course of action would be to replace the trucks on a scheduled basis. However, such a plan would accomplish little — benefits would show up in only a few areas.

Of course, the third course of action would be to simply replace the whole fleet. But, here is where the financial problem reared its head. A good share of the company's cash reserves were held in a local bank. Since the company was located in a small town, not only was it the main employer in the town, but it was also the main depositor in the local bank. If a large capital expenditure were made, cash would be taken away from the community.

If, however, the money were used for plant expansion or other projects that would employ local help, the whole community would benefit. While this problem would not be serious to those in larger urban areas, it would present quite a problem for the company's financial officers. Naturally, there are many other areas in which a company can use its financial reserves, such as acquisitions, sales promotions, etc. There is another point to consider — whenever you have to make an equipment expenditure, there are other managers or persons in the company who will think their programs or positions are equally justifiable.

The fourth course of action would be to replace the entire truck fleet, but not through an outright purchase. They could either rent or lease the fleet — then they could "have their cake and eat it too."

A quick analysis by most engineering managers would result in a decision to choose either of the first three methods, but never to choose the fourth. After all, you are paying interest to someone else and he is making money from you.

Actually, this is the problem found in many production personnel:

they do not understand corporate financing. Renting and leasing programs do have certain tax advantages. Then, by adding the profits from increased production through use of new equipment, such a program may possess enough positive features to convince management to approve your replacement program.

Cost of Ownership

The moment you buy a lift truck, it starts costing you money through insurance, taxes, and interest. To offset these charges and to show a profit out of the operation, the machine should be put to work as soon as you receive it.

There are two different types of costs incurred in owning equipment. The first cost figure is called fixed costs and consists of the insurance, interest, and taxes. The second is operating cost, such as the costs involved in keeping the machine running: maintenance, fuel, new tires, etc.

Probably the most important fixed cost item to consider is the depreciation schedule on the truck. As soon as the truck is delivered, its value starts declining because of its use, wear, weathering, and simply through the passage of time. This decline in value is deductible as an expense or as depreciation from the company's taxable income. The useful life of the truck is determined by its wear and obsolescence and a depreciation schedule can be set up for taxation purposes. Most accountants prefer to depreciate equipment at the fastest possible rate, which permits charging the largest costs against the machine while it is new and best able to carry these expenses.

There are three methods for determining depreciation. The *straight-line method* is probably the simplest and most familiar to engineers. This method calls for subtracting the salvage value of the truck from its cost, then dividing that figure by the number of years the truck is expected to be useful. As an example: a truck that costs $6,000 when new and that has a salvage value of $1,000 and a useful life of 5 years, would be depreciated at $1,000 a year. The method gives a uniform method for figuring costs and quite often is the method used for equipment justification. However, the method does lack a fast writeoff and most accounting departments use one of the other methods for taxation purposes.

86 | Lift Trucks

Another method is called the *declining-balance method*. With this method, the truck can be depreciated at a much higher rate during the first few years of its life, but no salvage value is allowed. In this method, depreciation is figured at twice the value of the straight-line method, except it is based on the balance of the truck's original cost at the beginning of each year, minus any depreciation already deducted. For example: again using the $6,000 truck with a useful life of 5 years, under the straight-line method its depreciation rate was determined to be 20% each year. With the declining-balance method, its depreciation rate would be 40% — only this method is figured on a yearly basis after the preceding year's depreciation is subtracted. At the end of the first year the truck's depreciation would be 40% of $6,000 or $2,400. At the end of the second year it would be 40% of $3,600 ($6,000 − $2,400) or $1,440. At the end of the third year the depreciation would be 40% of $2,160 ($3,600 − $1,400), and so on. At no time, however, can the entire cost of the machine be depreciated.

The third method is called *sum-of-the-years' digit*. This method allows a faster early-years' writeoff than the straight-line method, but not quite as fast as the declining-balance method. However, it does allow a complete writeoff. To figure the sum-of-the-years' depreciation is best shown by an example. Again using the truck cost of $6,000, 5 years useful life and $1,000 salvage value, we come up with a value of $5,000 ($6,000 − $1,000) to be depreciated over 5 years. The first step is to add the digits for the five-year period $(1 + 2 + 3 + 4 + 5)$, which equals 15. (If the period were 6 years, the sum would be $1 + 2 + \cdots + 6 = 21$.) Using 5 years as the allowable depreciation period, the first year's depreciation would be 5/15 of $5,000 or $1,667. The second year would be 4/15 of $5,000 or $1,333, the third year would be 3/15 of $5,000 or $1,000, etc.

In summary the rates of depreciation are as set forth in table 6-1.

The reason depreciation is so important is because of its effect on the company's income tax. Depreciation is charged to operating expenses and, as a result, reduces taxable income. The complexities of taxes cannot be taken lightly and it is advisable that you discuss the subject with your accounting department. As stated previously, in addition to the fixed costs one must also add the operating costs. Fuel and maintenance make up the biggest share of the operating expense; both will increase as the truck becomes older. Fuel usage increases to some

Table 6-1. Rates of depreciation by three methods

Year	Straight-line	Declining-balance	Sum-of-the-years
1	$1,000	$2,400.00	$1,667
2	1,000	1,440.00	1,333
3	1,000	864.00	1,000
4	1,000	518.40	667
5	1,000	311.04	333
Subtotal	$5,000	$5,533.44	$5,000
Sal. val.	$1,000	—	1,000
Total	$6,000	$5,533.44	$6,000

degree as the engine becomes less efficient through the years. Maintenance costs will make a fairly shallow curve at first and then will greatly accelerate as the truck becomes older. This is caused by the necessity of having to replace various components such as tires, batteries, and other parts.

Procurement Policies

New equipment is acquired usually because existing equipment either has reached the end of its useful life and needs to be replaced, or because it is required for a new production method.

Quite frequently, in order to obtain equipment for a new venture, a financial analysis is made through the use of the MAPI (the Machinery and Allied Products Institute) method or some other formula. The MAPI method is fairly easy to apply and one company's adaptation of this method is shown in figure 6-1. There have been many articles and books published regarding the MAPI method and detailed information on the subject is readily available from Machinery & Allied Products Institute, 120 So. LaSalle St., Chicago, Illinois.

Nevertheless, when it comes to lift trucks, most manufacturing managers and engineers are more interested in the replacement policy. On an average, the life of a gas truck is five years. However, depending upon the environment, maintenance costs, depreciation, etc., the truck may be good for ten or twelve years. As a result of these differences, when should you replace a truck?

Many lift truck dealers have charts and figures available that are al-

88 | Lift Trucks

```
XYZ Company                                         Sheet 1 of ____ Sheets
Estimate Number _____
Date _____
```

EQUIPMENT REPLACEMENT ANALYSIS

I. Description Present System (Dept. No._____) (Machine No._____)
   _____

   1. Salvage Value_____   2. Age_____

II. Description Proposed System_____
    _____

    3. Cost Installed_____   4. Service Life_____
    5. Estimated Salvage Value_____

III. Investment
    6. Installed Cost New Proposal (Line 3)................$_____
    7. Salvage Value Present Equipment (Line 1).............. _____
    8. Net Investment (Line 6 less Line 7).................. _____
    9. Reworking Cost of Present Equipment.................. _____
    10. Net Additional Investment (Line 8 less Line 9)....... _____

IV. Operating Advantage (Next Year)

| | Machine | | Proposed |
Advantages	Present	Reworked	Machine
11. Product Superiority..........$_____	$_____	$_____
12. Increased Output............			
13. Other Mfg. Factors..........			
Cost Advantages			
14. Direct Labor................			
15. Indirect Labor..............			
16. Machine Repair..............			
17. Tooling.....................			
18. Spoilage....................			
19. Down Time...................			
20. Floor Space.................			
21. Other (Taxes, Insurance, etc.)			
22. Totals......................	_____	_____	_____
23. After Tax Equivalent			
(Line 22 X (1 - Tax Rate))....$____(A)	$____(B)	$____(C)
24. Operating Advantage.......................$_____	$_____
	(B-A)	(C-A)

V. Return on Investment
   25. Yearly Capital Allowance
       (See additional sheet)....................$_____ $_____
   26. Next Year's Savings
       (Line 24 less Line 25)....................$_____ $_____

FIG. 6-1. A machine investment analysis form used for determining the economics involved in purchasing a new piece of equipment.

Financing | 89

ready calculated for almost any truck need. As an example, one company can easily set up a replacement schedule through the use of a computerized program. All one has to do is give them the basic data on his operation and a complete printout is readily available (fig. 6-2).

Another company has also developed a replacement program that can be used easily. They have charted the various obsolescence factors and one can calculate his own replacement program (fig. 6-3).

These particular formulations have proven successful for many plant managers and engineers. However, companies are beginning to question the need for replacement equipment since they could place their cash reserves in a financial institution and make a profit through investments. Therefore, most future replacement programs should take into account the time value of money in addition to the operating costs of the vehicles. Such a program is shown in table 6-2, which determines the useful life of a lift truck, while taking into account the appreciating value of money if that money had been invested elsewhere.

There are times when the use of a battery-operated truck is justified over the use of a gas-operated truck, and vice versa. When the physical requirements of the job dictate the choice of one truck over another, a

```
                                                                          0110000273

                                2000# GAS S            WITH- TRIPLEX           SHIFTS- 2
                                                                        OP.COND.- ABRASIVE
                                                                        TIRE REPL.- HEAVY

YEARS                 1       2       3       4       5       6       7       8       9      10
RUNNING HRS.       3,000   6,000   9,000  12,000  15,000  18,000  21,000  24,000  27,000  30,000

INIT. COST         5,335   5,335   5,335   5,335   5,335   5,335   5,335   5,335   5,335   5,335
LESS RESALE        1,907   1,387   1,214   1,041     867     763     694     624     555     486
NET COST           3,428   3,948   4,121   4,294   4,468   4,572   4,641   4,711   4,780   4,849

CUMULATIVE MAINT.
  PREVENTIVE MAINT. 571.65 1176.60 1831.20 2503.20 3225.75 3964.80 4755.30 5561.40 6419.85 7293.00
  ROUTINE REPAIR    759.72 1563.69 2411.92 3304.40 5174.20 6997.67 8121.60 10536.33 11802.92 14191.24
  OVERHAUL             .00     .00 1093.65 1123.75 2815.40 3173.00 4734.81 5374.74 7135.83 7721.78
  TIRES             455.16 1062.04 1668.92 2124.08 2730.96 3337.84 3944.72 4399.88 5006.76 5613.64
TOTAL MAINT.       1786.53 3802.33 7005.69 9055.43 13946.31 17473.31 21556.43 25872.35 30365.36 34819.66

TOTAL COST         5214.53 7750.33 11126.69 13349.43 18414.31 22045.31 26197.43 30583.35 35145.36 39668.66

MAINT.COST/HR        0.60    0.63    0.78    0.75    0.93    0.97    1.03    1.08    1.12    1.16

TOTAL COST/HR        1.738   1.292   1.236   1.112**  1.228   1.225**  1.247   1.274   1.302   1.322

* NOTE- ABOVE FIGURES FOR DEMONSTRATION PURPOSES ONLY. FOR ACTUAL VALUES TO
        BE EXPECTED IN YOUR APPLICATION, CONTACT YOUR YALE REPRESENTATIVE.
```

FIG. 6-2. A computerized printout showing that, in this particular example, the truck should be replaced after four years. Courtesy of Eaton Corp., Industrial Truck Div.

90 | Lift Trucks

```
                REPLACEMENT FORMULA FOR GAS AND LPG POWERED
                  RIDER TRUCKS UNDER 15,000 POUNDS CAPACITY

   MODEL:_____   CAPACITY:_____   RUNNING HOURS:_____

   DATE PURCHASED:_____    COST OF REPLACEMENT:_____

   1. Trade-in value                                              $_____

   2. Major repairs (overhaul, tire replacement, etc.)*            _____

   3. Unscheduled downtime costs (Table I)*                        _____

   4. Obsolescence costs (Table II)                                _____

   5. Estimated recovery or savings                               $_____

   6. Cost of replacement truck (new)                              _____

   7. Replacement advantage per cent line 5 ÷ line 6                        %

   8. Critical rating per cent prompting replacement (Use Table III)**      %

   *  Estimated for coming year.
   ** Where replacement advantage per cent exceeds critical rating per cent
      consider replacement.
```

Table I Unscheduled Downtime		Table II Obsolescence		Table III Critical Rating Percentage	
Running Hours range	Unscheduled downtime (as % of current replacement cost)	Truck Age	Obsolescence (as % of current replacement cost)	Running Hours range	Critical Rating (on current replacement cost)
0 - 2000	2 Per Cent	1	0 Per Cent	0 - 2000	Per Cent 90 or over
2 - 4000	4 " "	2	2 " "	2 - 4000	80 " "
4 - 6000	8 " "	3	5 " "	4 - 6000	70 " "
6 - 8000	10 " "	4	8 " "	6 - 8000	60 " "
8 - 10000	13 " "	5	11 " "	8 - 10000	50 " "
10 - 12000	18 " "	6	14 " "	10 - 12000	45 " "
12 - 14000	20 " "	7	18 " "	12 - 14000	35 " "
14 - 16000	24 " "	8	22 " "	14 - 16000	25 " "

FIG. 6-3. Replacement formula for calculating the useful life of gas powered industrial lift trucks from *Clark material handling symposium*. 1966. Battle Creek, Mich. Clark Equip. Co., Industrial Truck Div.

Table 6-2. Lift truck replacement schedule

Number	Item	Source and/or calculation	Year 1	2	3	4	5, etc.
1	Salvage value	Lift truck manufacturer or depreciation calculations					
2	Present worth factor	Any financial textbook. For 10% PW factor, see appendix.					
3	Present worth of salvage value	Item 1 times item 2					
4	First cost less present worth of salvage value	First cost is initial cost of truck. PW of SV is item 3.					
5	Capital recovery factor	Any financial textbook. See appendix for 10%.					
6	Annual cost of capital	Capital recovery factor times item 4.					
7	Operating cost	Truck maintenance records					
8	Present worth of operating cost	Item 2 times item 7					
9	Total present worth of operating cost	Cumulative total of item 8					
10	Average annual operating cost	Item 7 times item 9					
	Total average annual cost	Item 6 plus item 10					

financial analysis between the two is not necessary. When the trucks are competitive, accurate data to use in your analysis may be hard to come by. Charts and graphs furnished by trade associations, for example, may be somewhat slanted.

The cost analysis can center around three major areas. The initial investment may or may not include the battery and charger. For the first analysis, since batteries and chargers can now be leased or rented, it would be advisable to figure only the initial price of the truck and its depreciation schedule. Even though the electric truck costs more initially than the gas-operated truck, its estimated life is usually longer, resulting in lower cost on a yearly basis. Depending upon how your company writes off its equipment, this may not necessarily be any advantage.

The next area to be checked is the operating cost of both gas-powered and electric-powered trucks. This should be rather easy to calculate as one will mainly be comparing the fuel cost of one truck versus the other. Accurate records are required and these may be obtained by installing hour meters on the trucks and recording their respective fuel usage. Comparing use of gasoline to use of electrical energy may seem like comparing apples to oranges, but they all have one common denominator — the dollar. By knowing the amount of gasoline your trucks use and the number of hours they run per year, cost of the operation on a cents-per-hour basis can be determined.

The cost of operating electric-powered trucks can also be determined. The amount of electrical energy used for recharging the batteries, plus the average battery rental or lease fee on an hourly basis added to the amount of time the electrics are in operation, should produce a cents-per-hour figure.

Probably the most difficult area to calculate will be the maintenance cost factor. Maintenance on electric trucks will be lower than that on gas, but there should also be a battery maintenance cost added to the electric truck figure.

Actually, one should not say he is comparing gas trucks to electric trucks. The comparison is actually made of an internal combustion (IC) engine and an electric motor. The IC engine may operate on gas or propane fuel or the truck may be equipped with a diesel engine. The fuel used in an IC engine or the fact that a diesel engine may be used can greatly affect the maintenance cost factor. Therefore, you may have to acquire your own data since most charts on gas trucks versus electric trucks do not take into account propane fueled engines.

Rent, Lease, or Buy?

As mentioned previously, most engineers abhor the idea of renting or leasing industrial equipment. Yet the "front" office regularly rents or leases typewriters, copying machines, computers, and telephone services. Renting or leasing plans are not without pitfalls; they have their own particular advantages and disadvantages. What it all really amounts to is: "Don't put on the blinders." Investigate every possible means of acquiring the new equipment if it has been proven economically feasible through a layout and financial study.

There are certain advantages to owning your own equipment, especially if your company has the funds to invest. It would be advisable to buy equipment if:

1. instead of using company funds, money can be borrowed at reasonable rates;
2. the tax situation would be more favorable by an outright purchase;
3. the company has favorable financial and operating ratios;
4. the company has existing maintenance facilities that are efficiently run and, as a result, little downtime would occur in the operation;
5. overhead costs are low.

When a company does buy equipment, there is also the option of acquiring new or used equipment. There is probably some truth in the old saying, "When you buy something used, you buy someone else's problems," but there are also some good reasons for purchasing used equipment.

If a lift truck is to be used only part of the day, for emergencies, or as a standby for peak production periods, then the used truck can fit in quite well. Many used trucks have been thoroughly reconditioned and a limited warranty can be obtained as to their reliability.

However, if the operation requires that the truck be used constantly or if a new machine will operate more efficiently, then a new truck should be purchased.

A rental plan is another option often considered by companies. Renting a truck is ideal if it is required for a period of one year or less. If the company's business is seasonal, renting equipment is probably the best method, as renting eliminates depreciation or storage cost.

A rental plan can also eliminate maintenance worries. Maintenance is the responsibility of the supplier, but there is one drawback that should be pointed out. If the supplier is not located nearby, there may be considerable delays in getting a mechanic or a replacement if the rented vehicle fails.

Renting equipment offers tax advantages. A true rental program can be treated as an expense item for tax purposes and, therefore, the payments are deductible. The rental agreement conserves capital, which, in turn, provides additional sources of credit since your company will have operating capital instead of equipment capital.

One of the biggest advantages of renting is being able to test equipment under your own operating conditions. Most rental agreements contain an option to purchase and, if the truck performs satisfactorily, you may keep it.

Leasing programs usually have many of the same benefits as rental programs: freeing working capital, providing tax advantages, etc. A lease stipulates that the supplier pays the taxes and license fees and often also furnishes a maintenance service.

A lease also provides a method for accurately determining the costs associated with government programs. When a contract reimbursement is based on costs, the leasing or rental charges are readily available. If you own the vehicle it may be difficult to accurately determine such charges.

Hidden cost areas may show up when a plant maintains its own vehicles. Most maintenance men are not experts in maintaining lift trucks. Unless they have been thoroughly trained they may create problems with warranties on equipment you purchase. Such trouble can be avoided through the use of outside maintenance. In-house maintenance also requires the storage of spare parts. To this expense you must add the costs incurred in purchasing and receiving the parts (the average cost of processing the necessary paperwork and receiving the materials is estimated at $20.00 per order). The costs of the accounting department are not included: Checking of invoices, posting of books, sending out payments to several suppliers instead of one, etc.

A lease is not a cure-all for all problems; the best way to determine whether to rent, lease, or buy is to make a "cash flow" analysis. This can be done with the assistance of your financial department. A summary of factors to be considered in renting, leasing, and buying is shown in table 6-3.

One last word about obtaining new equipment: depending upon the situation, existing equipment can often be rebuilt or modified. Most lift truck dealers will rebuild your truck to a "like-new" condition. If there is no need for modernization and the usage is not heavy, then rebuilding might resolve your equipment problems.

Table 6-3. Comparison of advantages of renting, leasing, and buying a lift truck

Your position	Rent	Lease	Buy
1. You have adequate funds			X
2. You have reasonable borrowing power			X
3. You want to free capital	X	X	
4. You have low overhead			X
5. The tax situation is favorable to purchase			X
6. Your materials hndlg. is seasonal	X		
7. You want to conserve your credit rating	X	X	
8. You want to eliminate equipment obsolescence	X	X	
9. You have an excellent maintenance department			X
10. You want to eliminate maintenance	X		
11. You don't know how long you will need the equip.	X	X	
12. You want 100% financing		X	

7 | Driver Training and Safety

Often overlooked, but extremely important, is the training of the truck driver. Industrial trucks are not cars — they are sophisticated pieces of industrial machinery — and their operators need as good training as that received by lathe and milling machine operators.

Unfortunately, many plant supervisory personnel have not operated a lift truck. Their exposure to lift trucks has been through textbooks or through observations on the shop floor. They often believe that anyone who can drive a car can operate a lift truck and, as a result, the newest worker often finds himself on the truck. Lift truck operators are often the lowest paid of shop workers.

Plants that follow such a procedure are asking for trouble. A driver's license is just one of the prerequisites for becoming a lift truck operator. A truck operator must be physically fit and should have a thorough knowledge of the job.

More important, from an economic standpoint, the new Occupational Safety and Health Act of 1970 requires that you follow safe operating procedures in your plant. This law, which took effect April 28, 1971, covers all business engaged in interstate commerce (or about 4.1 million establishments and 57 million employees) and will affect almost every phase in manufacturing.

For example: A civil penalty of up to $1,000 must be assessed for each serious violation and may also be assessed for any nonserious violations. Willful or repeated violations may bring fines upwards of $10,000 and, should a willful violation result in an employee's death, up to six months in prison.

The new law may sound tough to plant supervisory personnel, but there are sound reasons for protecting the worker. On the average, every year sees 14,000 on-the-job deaths and more than 2 million disabling injuries in U.S. industry.

Under this law, the worker may request an inspection if he believes that a violation of job safety or health standards exists that threatens him in your plant. Furthermore, he may file a complaint with the Department of Labor if he feels that you discriminated against him or have discharged him for taking action under the act.

Regardless of the law, there are many reasons to be concerned about your lift truck operations. The following list contains some of the results of inefficient lift truck operations:

1. Product damage: A poor operator may dump a load, run his forks through a load, or he may place too heavy a load on top of another load. One damaged load may pay for a driver's complete training program.
2. Equipment damage: Most industrial trucks will cost at least $5,000 to $10,000. Truck downtime may be excessive due to the driver's carelessness. He may hit and ruin other production equipment with his truck.
3. Building damage: Many racks and building columns have been damaged by the rear-end swing of a lift truck. Too often this type of accident is not reported and the structure may collapse later with no one knowing why. In addition, a driver may hit an overhead utility line with his mast.
4. Personnel safety: The worker is your plant's most important commodity. One bad accident will pay for many, many driver training programs.

The lift truck may be a motor vehicle, but it is definitely not a family car. Although it is smaller than an automobile, the average lift truck weighs considerably more and could also carry the car on its forks. All this weight means that a lift truck has considerable momentum when it is moving.

In addition to driving a vehicle, a lift truck operator must also manipulate a load on the front end of the truck. He should have good depth perception and coordination for he will often store loads in high locations with only inches of clearance.

Knowledge of the job and equipment is as important to the truck driver as it is to a machinist first operating a lathe or milling machine. If the potential operator does not have the aptitude for the job, find out before he becomes an operator. Don't wait for an accident to occur and then pull him off the job.

How to Get a Training Program Started

First of all, don't try to make up your own training course. There is no need to print manuals or try to write rules and regulations. In addition to manuals offered by truck manufacturers, here are two training programs that can be purchased: *Powered lift trucks* by E. I. DuPont DeNemours & Co., Inc., Industrial Training Service, Room 7450 Nemours Bldg., Wilmington, Delaware 19898; and *Operating your fork lift truck safely* by Material Handling & Training System, P.O. Box 39094, Cincinnati, Ohio 45239.

Basically, all training programs should contain: (1) Classroom training and discussion; (2) actual field training on a truck; and (3) written and actual driver's testing.

The classroom portion of the program should cover the physical makeup of the truck and driving rules and regulations. Included in this portion should be a trip to the maintenance department to show the operator the importance of having a well-maintained truck. He is more apt to take care of the truck if he knows how much time and effort is required to make it safe and efficient for his operation. He should learn how to spot maintenance problems before they occur. He should also realize that having a good truck to drive is a team effort between him and the maintenance department.

Field training should be broken into two parts. We mentioned previously that a lift truck is driven as a vehicle and that it is also used to manipulate a load on its front end (working the truck). Therefore, the first training lesson should be learning how to drive the truck. An operator can get used to starting and stopping the truck, backing it, steering it, and picking up and lowering a pallet.

The next phase of field training is learning how to work the truck. This phase of training can include placing pallets beside one another, on top of one another, in racks, etc.

Last but not least should come the actual driver's test. Just as a man takes a written and driving test to obtain his license to drive a car, the same holds true for a lift truck. As a result of a training program, only qualified drivers should be allowed to use lift trucks in your plant.

Truck Safety

Up to this point we have been talking about initiating a safety program. Let's look at the truck and analyze some of the problems one can have because of the difference between it and a car.

One of the most important points to remember is that the lift truck carries its load on one side of the front axle with the weight of the truck counterbalancing the load on the other side of the axle. As long as the counterbalancing action of the truck is greater than the load, the truck will remain stable. However, if the load becomes greater than this counterbalancing action, the truck will tip forward.

What causes this action?

It should be noticed that the capacity of the truck is stated in so much weight at a load center of so many inches (for example, 2,000 lbs at 24 in., which means the maximum capacity of the truck is 2,000 lbs with its center of gravity 24 inches from the center line of the front axle). As a result, all the weight times the distance from the front axle must remain equal to or less than the stated capacity or the truck will tip forward. (From our example: What is the maximum weight the truck could carry if the load center of gravity was 36 in. from the front axle center line? W times 36 must equal 2,000 times 24. Therefore, W equals 1,333 lbs.)

Since the mast can tilt forward and the center line of the load can shift, it is always best to carry a load slightly less than the maximum allowable amount.

The following list includes some of the rules that should be taught to your truck operators:

1. When traveling down inclines with a load, it is best to drive down in reverse. If the operator drives down forward, gravity works against him. If the load is at all unstable, you can be assured that it is going to spill off the truck.
2. The truck should be used for carrying the operator only — no passengers. If the truck is to be used for lifting personnel, a platform with a railing around it and one that is affixed to the forks should

100 | Lift Trucks

Fig. 7-1. Montage of driver training illustrations shows the many phases of a driver training program. Humor and instruction, as well as actual driver tests, help the operator become proficient. Courtesy of Clark Equipment Co., Industrial Truck Div.; and White Materials Handling Division, White Motor Corp.

Driver Training and Safety | 101

be used. Under no conditions should a man ride the forks up or ride up on a pallet or skid.

3. Never overload a lift truck by placing weights on the back of the truck, which in turn would allow a greater load to be lifted on the front end. Granted, the counterbalancing effect would be increased, but it must be remembered that the hydraulic and mechanical components of the truck were not designed for the higher loads.
4. Loads should always be transported low on the forks. Carrying a load high increases the possibility of tipping the truck over, especially if the truck must be brought to a sudden stop.
5. The forks should touch the floor when the truck is stopped. This will prevent someone from walking into them. It would be especially dangerous to leave the truck parked with the forks at eye or head level. When parking a truck be sure to shut the motor off and remove the keys.
6. Never allow an unauthorized person to use a truck.
7. Make sure the floor-loading capacity of the building will support the weight of the truck. Many old mill buildings will not support the weight of a lift truck and load.
8. Insist that the driver be especially careful when loading over-the-road trucks and railcars. If dockboards are not attached to the building, be sure that approved portable dockboards are used. A steel plate is a poor substitute for a dockboard. The driver should also check carefully to be sure the wheels of the trailer he is loading have been chocked. Driving a lift truck into a trailer and then braking will push the trailer away from the dock unless it is blocked.
9. Many lift trucks were not built for towing. If the truck does not have a towing coupling, it probably wasn't designed for towing.
10. The driver should watch out for low clearances.
11. Loads should be carried with the mast tilted back toward the driver. If the load is so high that forward visibility is impaired, the truck should be driven in reverse.
12. When driving across railroad tracks, always drive diagonally. Tracks should never be hit straight on.

These are only a few of the more important rules for the operator. The following list includes rules that apply to you, the supervisor:

102 | Lift Trucks

1. Select the proper truck for the application.
2. Enforce driver training and development.
3. Remind foremen to keep aisles in their departments clear of foreign objects. Visibility in a lift truck is difficult under the best conditions and cluttered aisles are a real hazard.
4. Equip the truck with proper safety devices and then keep it in good repair. Overhead guards are an absolute must in many states and just plain good sense in any state. Other safety devices may include warning lights, horn, and a fire extinguisher.
5. During any new construction project, make sure that doorways are large enough for the trucks to pass through, install ramps from one floor level to another, specify permanently attached dock boards, etc.

Lift trucks are important pieces of machinery in the proper flow of materials through the shop and supervisors should be kept aware of this fact. Any short cuts, such as using a truck with faulty brakes instead of taking it to maintenance for repair, is a dangerous procedure that cannot be tolerated.

In addition to the previously mentioned rules for driver and supervisor, there are some rules that should be observed by everyone.

1. Gas trucks should be used only in areas with adequate ventilation.
2. Racing the engine or driving with a faulty muffler is bad practice — both create noise and noise is a form of pollution.
3. Gas or LP fuel trucks are a fire hazard and no smoking is the rule when fueling these trucks.
4. The operator needs the cooperation of his fellow workers. Pedestrians should be reminded that although the operator will give them the right of way, he may not always be able to see them, so it is better to be safe than sorry. In any contest, the weight and momentum of a lift truck will always make a loser out of the pedestrian.
5. Watch the loads on the forks. Although it is assumed that the operator has a secure load, keep away from it. Above all, never stand under the forks of a lift truck.

Driver training and education start right at the top of the management level. If you are part of management, encourage your subordinates to support a good driver training program. If you are further down the ladder, list the values of driver training and enlist the help of your superiors to support your program.

Driver training can be good for plant public relations. Driver training programs can be covered by the company newspaper. Safety awards can be set up for good driving and by your active interest employee-employer relations can go only one way — up.

Bibliography

White Mobilift. *Lift truck pilot training manual.*
White Mobilift. *Lift truck safety as response-ability.*
Eaton Corp., Industrial Truck Div. *Instructor's Manual; industrial truck operators' training program.*
Towmotor Corp. *The color of danger.*
Clark Equipment Co. *The professionals; tips on safe fork truck driving.*
1964. Play safe with lift trucks — Follow these rules. *Plant Eng.* Dec. P. 131.
Clark Equipment Co. *In-plant driver training.*
1971. Processors: Safety act will nudge you into progress — Ready or not! *Plastics World*, Sept.
1971. Safety — What the new law demands of you. *Mod. Materials Handling*, Sept.

8 | Maintenance

Downtime is the dread of all production shops and the lift truck needs proper maintenance. In a smaller plant, the truck may be the heart of the plant's materials handling system. If the truck is down, the entire work flow in the shop may come to a halt. Tips on good maintenance as well as the facts of obsolescence are given in this chapter. A proper maintenance program can make the difference between an efficient operation and one plagued with excessive amounts of downtime.

In a study by the Clark Equipment Co., the fact was brought out that a lift truck's productivity is affected more by its reliability than by the truck's speed. The study indicated that many factors were involved in decreasing a truck's utilization of speed. Unstable loads, pedestrians, stopping for other trucks, turning corners, backing down ramps, etc, all prevent a lift truck from working at its maximum speed. In fact, these factors can decrease a truck's speed by 30 percent and, sometimes, even more. Since many of these items were beyond the control of the driver, excessive downtime was the difference between a productive truck and a nonproductive truck. As a result, reliability was far more critical than had previously been realized.

Preventing Downtime

The key to preventing excessive downtime is keeping the lift truck fleet well maintained. First step in setting up a maintenance program is to decide whether to:

1. use contract maintenance;
2. use an in-plant work force for maintenance; or
3. use a combination of 1 and 2.

Each method has its own merits and its success depends upon the competence of the plant's maintenance department and the size of the plant. It is easier for large companies to have their own in-house truck maintenance program, as they usually have the facilities, experience, and personnel for such an operation.

Small companies should probably have a contract maintenance program. Usually, their maintenance departments are understaffed and the only maintenance given is when a truck is actually out of order. In these cases contract maintenance will insure a preventive maintenance program.

I believe a combination program offers the best solution to the average plant. The truck vendor sends his man in once a month to lubricate the trucks and make necessary repairs. The plant's mechanic performs any daily and weekly maintenance checks and does the troubleshooting.

Records

Records are most important in the development of any maintenance program. They provide the data for determining parts replacement and maintenance checks as well as determining when a truck has reached the end of its economic life. Too often, trucks are costing their owners more than the price of a new one and should be replaced.

Records should be simple, yet complete (fig. 8-1). Once the records have been started, they should be kept up to date and accurate. Each truck should have its own assigned number and hour meters should be installed on them to provide an accurate record of how long the truck is actually being used.

A planned maintenance program may not end all breakdowns, but it will keep unscheduled downtime to a minimum.

One area of truck maintenance that requires particular attention is the safety check: brakes, hydraulic system, warning lights, horn, and all the other items that involve the prevention of property damage or personal injury. All safety items should be checked daily and, if any prob-

Checklist

```
                                    Date_____
Make_____ Capacity_____ Serial No._____
Hour Meter_____ Plant_____ Department_____
Truck Number_____ Mechanic_____ Attachments_____
Type of Inspection:   8 Hours ☐   100 Hours ☐   500
Hours ☐   1000 Hours ☐
```

SECTION OF TRUCK	A	B	C	D	REMARKS
1. UPRIGHT (Cyl., Brkts., Slide Strip, Piston Head, Bearings.)					
2. TILT CYLINDERS (Packings, Boots, Leathers, Pins, Bushings.)					
3. DRIVE AXLE (Diff., Seals, Bearings, Shaft, Gears.)					
4. BRAKES (Shoes, Cylinders, Lines, Master Cylinder.)					
5. TRANSMISSION (Levers, Gears, Seals, Controls.)					
6. HYDRAULIC SYSTEM (Pump Valve, Lines, Sump Tank.)					
7. PEDALS, FRAME, GAS TANK FLOOR BOARDS.					
8. CLUTCH (Throw out Bearing, Cover, Driven Plate.)					
9. RADIATOR, OIL FILTER, AIR CLEANER, MUFFLER.					
10. ENGINE (Engine Accessories, Motor Supports.)					
11. TIRES, WHEELS, SEATS.					
12. STEAMCLEAN ☐ PAINT ☐					
13. STEERING GEAR (Hand Wheel, Drag Link.)					
14. STEERING AXLE (Spindles, Steering Arm, Tie Rod.)					
15. POWER STEERING					
16. ELECTRICAL SYSTEM (Voltage Regulator, Gauges, Horn, Battery, Wiring.)					
17. ATTACHMENTS & OPTIONAL EQUIPMENT					
18. MISCELLANEOUS					

```
A. Inspect    C. Clean      Signature
B. Lube       D. Repair
```

FIG. 8-1. Records on lift trucks should be complete, but need not be complicated, as this checklist shows. Courtesy of Clark Equipment Co., Industrial Truck Div.

lems are present, they should be corrected immediately. Trucks with faulty safety gear should not be used under any conditions.

Internal combustion engines also require safety checks. Gas-line leaks, noisy or leaky mufflers, or other problems that could cause pollution problems should be corrected immediately. For example, many gas-powered trucks are now equipped with catalytic exhaust purifiers. If leaded gas is used in the plant, the catalyst agent will probably have to be replaced every 500 hours. If the truck uses LP gas or unleaded gas, the change will be required every 2,000 hours. Always follow manufacturer's specifications and recomendations when using catalytic purifiers.

A word of caution: the catalyst may not need changing — the trouble may be elsewhere. The engine may be burning oil and it should be checked for piston ring wear or for sticky valves. In either case, the purifier can be damaged if the problem isn't corrected immediately.

An improperly tuned engine is a potential pollutant and, with the present emphasis on pollution control, workers will no longer tolerate working in a dirty environment. Noise pollution is equally important, and a leaky muffler is usually a noisy one — replace it immediately.

Truck maintenance differs depending upon whether the truck is gas- or electrical-powered. If the maintenance department is assuming complete responsibility for truck maintenance, then a program should be instigated to send mechanics to a manufacturer's training school. This allows a plant's mechanic to learn the proper techniques of truck maintenance.

Gas Truck Maintenance

The following items should be checked on trucks during safety checks: horn, lights, and brakes. The oil level should be checked on both the engine crankcase and the hydraulic oil sump. All warning lights on the dashboard should be checked; also, oil pressure, radiator temperature, etc. The tires should be inspected for cuts, alignment, and security of mounting. The forks should be tested by lifting them to full height and tilting them fully forward and backward.

Normal maintenance checks are usually carried out on a 100, 500, and 1,000 hour basis. This is why it is so important to have an hour meter on a lift truck.

Usually, a good lubrication program is the key to a good lift truck PM program (fig. 8-2). Most other maintenance checks can be made when the truck is brought in to be lubricated. Every point that can wear or become damaged is lubricated and can be noticed during the lube program.

Techniques are available to check the longevity of motor oil. Oil is changed in most trucks periodically; however, some trucks are operated in a dirty atmosphere and should have their oil changed more fre-

108 | Lift Trucks

Fig. 8-2. Good lubrication is a must for any lift truck. Adequate equipment, such as this truck lift, helps the mechanic reach all parts of the truck. Courtesy of the Marmac Co., 1231 Bellbrook, Xenia, Ohio 45385

quently. Ask your lubricant supplier or lift truck dealer for proper lube requirements.

Clean oil extends the engine life as well as the life of your hydraulic system. In addition to oil changes, the filter elements should also be changed periodically. However, the oil should be changed any time dirt appears, discoloration occurs due to high operating temperatures, or a strong odor is detected. If metal particles appear in the oil, some truck part has probably failed. Always use clean oils and clean lubricants.

Maintenance | 109

Sludge is another problem; it is formed when the engine is operated at low temperature. In the combustion process, water, sulfuric acid, sulfurous acid, and many other chemical compositions are formed when the fuel and air burn. These compositions are gaseous and will condense on cold surfaces (about 135°F or lower). Therefore, the crankcase temperature should be kept above 135°F.

Sludge will usually form in the valve spring chamber and cause corrosive wear on tappet faces, rings, and valve stems. In addition, sludge can cause carboning of the rings and it can plug oil passages and destroy seals.

In cold climates, before placing an engine under full load, let the engine warm up by running it slowly; if there is any evidence of sludge, change the oil.

All engines are different. A diesel will have a different fuel system than a regular gas engine. Cleanliness is the most important consideration in keeping any fuel system in working order. The fuel supply area should be kept clean and all safety procedures should be followed when fueling or cleaning the fuel system of a truck.

In a gas engine, the electrical system and fuel system must work together. The fuel system consists of a tank, pump, filter, carburetor, intake manifold, throttle, and connecting fuel lines. The electrical system is made up of a battery, generator (or alternator), distributor, starting motor, coil, regulator, condenser, spark plugs, and ignition switch. Basically, the system works as follows: The ignition is turned on, the battery provides the power for the starting motor to turn the crankshaft of the engine. The engine also powers the fuel pump which draws fuel from the tank and forces it into the carburetor where the fuel is vaporized and mixed with air. The vaporized mixture then enters the engine through the manifold. The spark plug ignites the mixture and the engine starts. Spark is provided by taking low voltage from the battery, sending it through the coil where it is stepped up several thousand volts, and then the distributor sends the current to the proper cylinder at the proper time.

The generator converts mechanical power of the engine into current and sends a flow of current to the battery. The regulator controls the current generated so that the battery won't be overcharged.

There are a few maintenance requirements in the electrical system and, although not too difficult to perform, they are quite important for

110 | Lift Trucks

the most efficient operation. Spark plugs should be cleaned periodically and the gap should be adjusted to manufacturer's specifications. Burned or cracked spark plugs should be replaced.

The distributor controls the current to the spark plug so that the fuel will ignite at the most efficient point at the end of the compression stroke of the piston. This requirement varies according to the type of fuel used and the design of the engine. Serious damage can be done to the piston ring and valve damage can occur if the timing is off (fig. 8-3). Therefore, timing should be done with a "strobe" or timing light and never by "ear." Correct ignition timing can never be stressed too

Spark Advance Characteristics

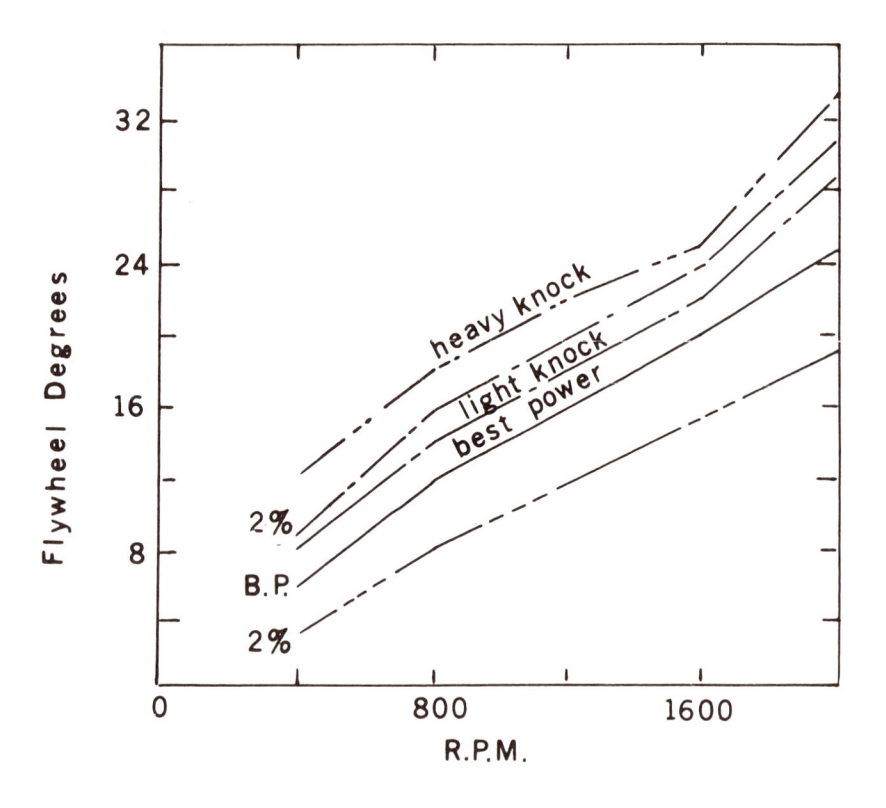

FIG. 8-3. Proper timing is very important as engine damage can occur if it is off. Courtesy of Continental Motors

Maintenance | 111

highly and, since this timing is based on the fuel used, correct octane rating of fuel should always be used.

Previously, it was mentioned that a cold engine can cause sludge to form. However, engines are provided with a cooling system to prevent metal temperatures in the engine from becoming too high. An engine runs best at a temperature range of 170° to 180°F. At this temperature, cylinder wear is less, fuel consumption is most economical, and the engine is operating at the best power output.

Air or water can be used as a cooling medium although most lift trucks are water cooled. The cooling system consists of a radiator, water pump, thermostat, and fan. The fan blows air across the radiator fins, where heat transfers from the hot water to the cooler air. The water pump is used to circulate the cool water from the radiator to water jackets in the engine. The water picks up the heat from the engine and then flows back to the radiator where it is cooled again. The thermostat is used to control the temperature of the water.

Soft water should be used in the radiator and antifreeze should be added if the truck is used in below-freezing climates. Usually a rust inhibitor is also added to the water.

If the thermostat valve remains in an open position, too much cooling water will flow through the engine and it will have a low temperature reading. If the valve remains in a closed position, the engine will overheat and cause water to boil over from the radiator. If the thermostat fails, it should be replaced.

The thermostat can be inspected by placing it in heated water and checking it to see that it opens at the proper temperature range. Many radiators are equipped with a pressure cap to improve the cooling capacity of the radiator. When a pressure cap is used, the system should be airtight. Because of the added pressure, hoses and connections should be checked frequently.

Fan belt tension is important — the tension should be neither too loose nor too tight.

The battery of the gas truck also needs additional water periodically; only distilled water should be added. More details on battery maintenance are covered in the electric lift truck maintenance section.

Electric storage batteries emit hydrogen gas when they are being charged and may continue to do so for some time after the charge is completed. Since hydrogen gas is highly inflammable, electric sparks or flame should be kept away from a newly charged battery.

The power plant transmission differential system in a gas truck is not quite as simple as that in an electric truck. However, a proper maintenance program should minimize any downtime problems.

Because of the variety of transmissions available, a booklet on maintenance procedures should be obtained from the manufacturer. The type of service and operating conditions will determine the maintenance schedule. However, the oil level should be checked periodically and, at the same time, checks should be made for any oil leaks.

In general, the following service procedures should be followed:

1. Stop engine before checking or adding oil.
2. Clean around the fill opening before adding or checking oil.
3. Always use clean oil and clean containers.
4. Change oil according to the manufacturer's recommendations or if the oil shows traces of dirt.
5. Drain dirty oil while the unit is warm.
6. Avoid overfilling the oil container.

Last but not least, one should always remember that dirt is the enemy of all moving parts. All filters should be cleaned according to instructions and maintenance work should be performed in clean environment. Trucks operating in a dirty atmosphere should have more frequent maintenance inspections. A summary of gas truck maintenance is shown in the following outline.

Gas-powered lift truck maintenance program

Every 8 hours or one shift

    Engine crankcase — Oil level check.
    Engine cooling — Radiator coolant level check.
    Engine air cleaner — Service complete cleaner assembly.
    Tires — Inspect for cuts and alignment. Inspect for poor inflation and remove any materials embedded in tires.
    Fuel tank — Check to see if tank is full.
    Lights — Visual check.
    Horn — Be sure it works.
    Brake pedal — Check for proper play, for drift or spongy pedal.
    Brake operation, parking — See if it holds on reasonable grade.
    Instrument panel indicators — Check oil pressure, water temperature, ammeter.

Maintenance | 113

Clutch pedal — Should have some free play.
Warning lights — Check all lights for engine oil pressure, converter oil, converter temperature, transmission oil pressure, and temperature as well as the engine temperature.
Lift and tilt operation — Raise to full height and tilt backward and forward. Check oil sump level.

Every 100 hours

Engine crankcase — Drain and fill.
Engine breather — Clean or replace.
Engine oil filter — Replace filter element.
Fan and generator belts — Inspect and adjust.
Generator — Lubricate.
Starter — Lubricate.
Fan pulley — Lubricate.
Water pump — Lubricate.
Steering gear — Check lubricant.
Battery — Check fluid level, connections, and take hydrometer reading.
Fluid coupling — Check fluid level.
Lift and tilt cylinders — Check for drift and leaks. Look at mountings.
Control linkage — Inspect and lubricate.
Hydraulic sump tank breather — Clean or replace.
Governor slip tube — Lubricate.
Transmission — Check oil level.
Transmission and converter — Check oil level.
Drop gear case — Verify level.
Differential — Check lubricant level.
Drive axle air vent — Service and clean.
Axle ends — Lubricate.
Axle end air vent — Service and clean.
Fuel lines and cap — Inspect.
Suspension — Check mountings.
Clutch pedal adjustment—Check and adjust.
Brake pedal — Adjust free travel.
Brake system — Verify fluid level. Check cap vent hole.
Steer valve and pump — Check for leakage.
Hydraulic control valve and lines — Inspect.
Lift bracket and piston head slide guides — Inspect and lubricate, check for wear.
Lift chains — Adjust and lubricate.
Visually inspect all hydraulic lines and electrical wiring.

Every 500 hours

Steam clean machine
Standard transmission — Drain and refill. Check air vent.
Automatic transmission — Drain and refill. Clean screen and replace filter.
Fuel filter — Inspect and clean.
Fuel pump strainer — Remove and clean.

114 | Lift Trucks

Hydraulic sump tank — Drain, clean and refill.
Hydraulic sump tank oil filter — Replace.
Steer gear — Adjust thrust and lash.
Steer axle linkage — Check and adjust turning radius.
U-joints—Inspect.
Exhaust system — Check for leaks.
Fuel pump — Clean pump bowl and strainer.
Muffler — Check for leaks.

Every 1,000 hours

Complete engine tuneup
Fuel pump — Check for proper operation.
Fuel pump strainer — Disassemble and clean.
Fuel sediment bowl — Clean.
Standard transmission — Drain and refill.
Automatic transmission — Make pressure check.
Fluid coupling — Drain and refill.
Generator — Check brush spring tension, replace if worn. Check commutator.
Voltage regular — Check.
Differential — Drain and refill.
Steer wheel bearings — Adjust, clean, and repack.
Brake system — Test, bleed, and fill reservoir.
Hand brake — Test.
Cooling system — Flush system, check, and refill.
Steer system — Check hydraulic pressures.
Main hydraulic system — Check pressures.
Drop gear case — Drain and refill.
Starting motor — Check brushes and for dirty commutator.
Hydraulic sump tanks — Drain and refill.

Electric Truck Maintenance

While an electric truck will probably require a lighter maintenance program than a gas truck, the power source (the storage battery) does require an extensive and conscientious maintenance program (fig. 8-4). Batteries should be assigned specific numbers to promote accurate recordkeeping.

Remember, the storage battery does not store electricity. The battery works in the following manner: When an external circuit is closed on the terminals of a charged battery, a chemical reaction takes place inside the battery. This reaction causes a current to flow through the battery. The solution in the battery is called an electrolyte and it is very acidic. Maintenance workers responsible for battery care should

FIG. 8.4. The battery is probably the most critical maintenance element in an electric lift truck. Workers should be very cautious when working around batteries as the electrolyte is very acidic. Courtesy of Exide Div., Electric Storage Battery Co.

wear rubber gloves, apron, and goggles and should be very careful not to spill any solution on themselves or on the equipment. Should acid be accidentally spilled, it can be neutralized with ammonia or soda solution. If acid should get in worker's eyes, wash the area with water and contact a doctor immediately.

These precautions are given, not because the storage battery is a dangerous item, but, like anything else, its misuse usually will create problems.

Industrial lift truck storage batteries are expensive and workers should be instructed to treat them with as much care as they would any other equipment. Proper care will insure maximum life.

As mentioned previously, a chemical reaction takes place inside the battery to generate current. In order to determine how much of a chemical reaction has taken place, the specific gravity (a solution's relative density as compared with water) of the electrolyte should be checked periodically with a hydrometer (fig. 8–5).

A hydrometer consists of a delicately balanced graduated glass float that rests in a syringe tube. Fluid to be measured is drawn into the syringe and the fluid's specific gravity is registered on the scale on the float. Since the hydrometer is based on a reference point of 77° F, the temperature of the battery should always be taken and the hydrometer reading should be corrected to a normal temperature reference point.

In order to recharge a battery after discharge, direct current is passed through the cells until the acid goes back into solution. There are various ways of charging a battery and the manufacturer should be consulted for his recommendation.

Some of the rules of good battery maintenance include:

1. Proper charging – this is most important for efficient operation. Overcharging the battery will shorten its life by causing excessive gassing, high temperatures, and wasted power. A battery in a discharged condition can be equally damaging and should be recharged immediately. A discharged battery will not function in cold weather as well as a charged battery and freezing of the cells may occur.
2. Cleanliness is equally important to a battery's life. Corroded or dirty terminals should be cleaned and any acid spilled should be promptly neutralized. Vent plugs should be kept tight and securely in place.
3. Height of the electrolyte is another factor which should be checked daily. Use only water which is approved for battery usage – such as distilled. The electrolyte level should never fall below the top of the plates and, when being filled, fluid should never be allowed above the vent well. Excessive water requirements probably indicate that the battery is overcharging.

In cases where a maintenance man mixes the sulfuric acid for a battery, be sure that he thoroughly understands the procedures for the use of acid. When acid is diluted with water, heat is generated. Therefore, the mixture should cool to at least 100 degrees F before it is added

Maintenance | 117

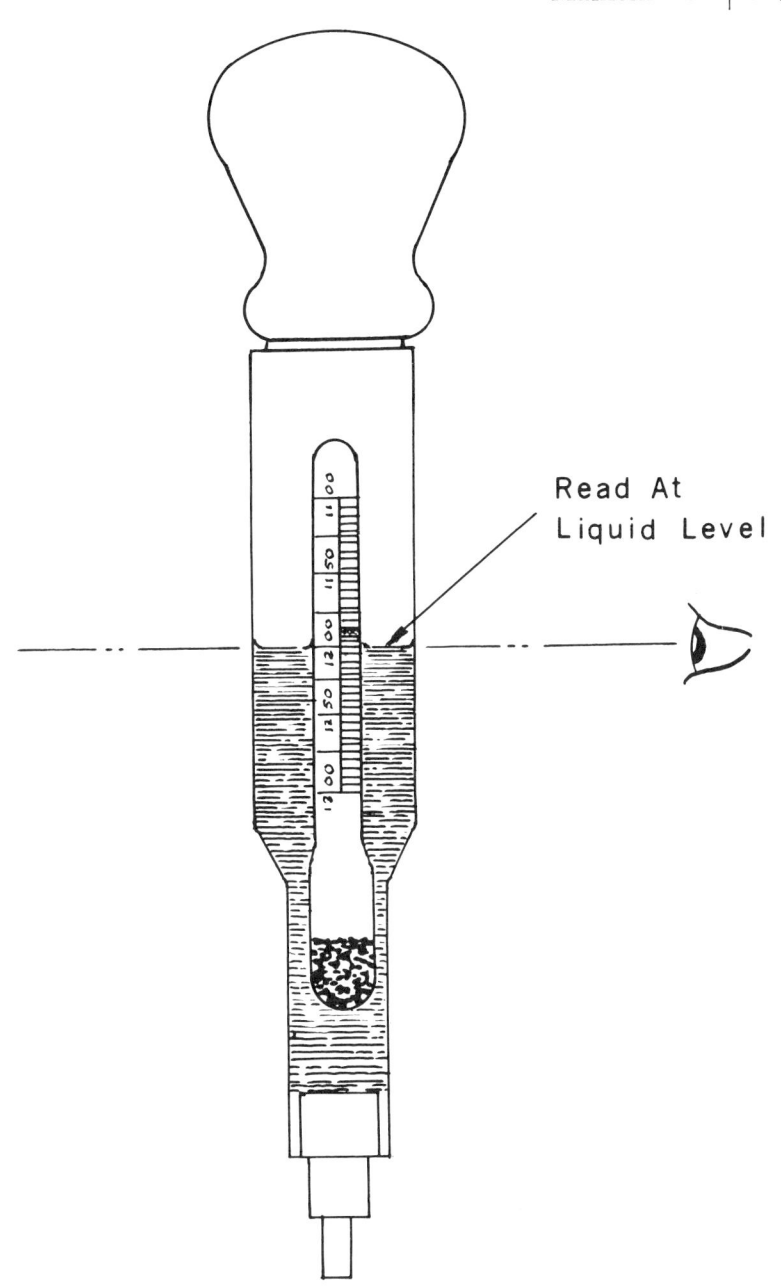

FIG. 8-5. A hydrometer is used to determine the specific gravity of a battery's electrolyte. For correct reading, the eye should be on a level with the surface of the liquid. Disregard the surface tension curvature (meniscus).

to the battery. Another word of caution: Always add acid to water — never add water to acid.

If the specific gravity of a fully charged cell is above normal, the electrolyte should be adjusted by removing some of the electrolyte and adding water until proper specific gravity is reached. Too high a specific gravity will ruin a battery.

If only a few electric powered trucks are in use, a contract maintenance program would probably prove beneficial. When the mechanic lubricates the truck he can also run a check on the battery. If extensive battery maintenance, such as making repairs, is contemplated, the maintenance mechanic should be given proper training by the battery vendor. A summary of electrical truck checks is given in the outline that follows.

Electric-powered lift truck maintenance program

Every 8 hours or one shift

    Set brake — Inspect.
    Battery — Check liquid level, take hydrometer readings. Inspect.
    Battery charge indicator — Check.
    Lights — Check.
    Horn — Check.
    Brake pedal — Check for proper free play. Check for drift or spongy action.
    Tires — Inspect for cuts and alignment.
    Hydraulic sump tank — Check fluid level.
    Lift and tilt operation — Raise to full height and tilt backward and forward. Check oil sump level.

Every 100 hours

    Contactor panel — Check and adjust.
    Electrical wiring — Check connections for tightness.
    Steering gear — Check lubricant.
    Differential — Check lubricant.
    Differential air vent — Clean.
    Brake pedal — Adjust.
    Brake system — Check fluid level.
    Visually inspect all hydraulic lines and fittings; steer valve and pump, hydraulic control valves, lift bracket and piston head guide. Lubricate.
    Lift chains — Adjust and lubricate.
    Lift and tilt cylinders — Check for drift and leakage. Check mountings.
    Hydraulic sump tank breather — Clean.
    Lubricate all other lube locations.

Every 500 hours

    Tires — Check.
    Hydraulic sump tank — Drain and refill.
    Hydraulic sump tank oil filter — Replace filter element.
    Steering gear — Adjust thrust and lash.
    Steer axle linkage — Adjust.
    Nuts, bolts, and cap screws — Tighten as required.

Every 1,000 hours

    Pump and drive motor brushes — Check brush spring tension and bearings and lubricate.
    Differential — Drain and refill.
    Hand brake — Test.

Other Maintenance Areas

In addition to the more ordinary areas of truck maintenance — hydraulics, engine, electrical system, etc — there are other areas in a truck's makeup that require a sound maintenance program and, yet, these areas are seldom given consideration.

Attachments

Most attachments are mechanically or hydraulically operated, so they also require maintenance. Rubber facings on carton clamps, chipped or rolled-up tips on push-pull forks, and leaking hydraulic fittings may require maintenance, yet are easily overlooked by the maintenance mechanic.

Since the hydraulic system would probably be the most critical item, the hose lines to the attachment should be checked at the beginning of each shift. If any damage is spotted, repairs should be made immediately, as a hose line break while the truck is running could cause a load to be dropped.

Rubber facing on a carton clamp or the facing on a vacuum clamp should also be checked for damage at the beginning of each shift. Loads may be torn by the damaged clamp, or, worse, the load may be dropped. Or misuse may also damage the carton clamp: such as pushing skids with the attachment, or opening or closing a railcar door with the clamp. The following is a daily maintenance checklist for lift truck handling attachments.

120 | Lift Trucks

Item	Look For
Hydraulic lines	Frays, gouges, or grooving connections
Hydraulic fluid	Leaking lines
Pressure control valve	Missing or improperly set connections
Clamp pads	Torn or loose rubber facing
Clamp arms	Arms out of parallel or damaged
Load back rest	Bending or breaking
Fork tips, pushing and pulling devices	Chipping, rolling, or splintering
Pusher	Face out of alignment
Pusher cylinder	Its connection to the truck

Tires

As with a car, safety of the industrial truck depends on its tires. All tires require maintenance checks, and, by selecting the right tires and insisting that the operator exercise care in the use of the truck, they need not become a maintenance problem.

With the many varieties of tires available, it is most important to select the right tire for the right job. There are pneumatic tires and solid rubber tires. Usually, trucks used indoors will be fitted with solid rubber tires. The tires may be constructed with or without tread, or they may be grooved. They may be the oil-resistant, nonmarking type and made of special materials such as reinforced steel and urethane.

Solid rubber tires are press-fitted onto a wheel and one can usually figure the pressure to equal 5,000 lbs per inch of wheel diameter. If a tire is 15 inches in diameter, then the press size would be 15 times 5,000 or 75,000 lbs ($32\frac{1}{2}$ tons). A tire can be replaced by pressing off the old tire while pressing on the new one. Pressure should always be applied through the metal band, never with rubber-on-rubber contact.

The condition of the tire may also indicate the existence of other truck maintenance problems. A tire that is worn more on one side than on the other may be evidence of a misaligned steering arm or perhaps a faulty axle.

The following list gives some of the procedures to follow to assure maximum life out of the truck's tires:

Maintenance

1. Tires should be inspected at least once each week.
2. Tires should be centered on the wheels. Axle alignment and steering should be checked periodically.
3. Aisles should be kept clean and free of obstacles.
4. Vehicles should be lubricated properly. Overlubrication can be harmful, as any grease or oil spilled is harmful to most rubber products.
5. Sufficient tire clearance should be allowed. It is recommended that there be at least one-half inch clearance between the tires and the truck frame.
6. Aisles should be clearly marked — this will help the operator avoid hitting curbs, posts, etc.

In addition to proper maintenance, a few words of advice to the operator can help in preserving the truck's tires. First, the proper equipment should be used in lifting loads. Not only will overloading put a strain on the lifting mechanism, but it can be equally harmful to the tires. Sharp turns, quick stops and starts, driving through oil or acids, and spinning the tires in metal chips or stones will quickly shorten the tire life.

Building Maintenance

There are four elements in any lift truck operation: truck, operator, load, and facility. If lighting is poor, or aisles too narrow, the truck cannot operate as efficiently as it should. Most important are the facility's floors. This is the point where the truck comes into contact with the facility and any defects in the floor will have an effect on the life of the tires. In addition, poorly maintained floors are a safety hazard to both the operator and the load.

Summary

In summary, seldom if ever is equipment overmaintained. Yet proper maintenance may make the difference between a break-even operation and one that shows a profit. With accurate record keeping and a planned maintenance program, you will have a more efficient lift truck operation and also a safer one.

Bibliography

1970. MMH maintenance guide. *Mod. Materials Handling* April.

Woodcock, Walter J. 1968. All-around maintenance keeps lift trucks on the job. *Materials Handling Eng.* June.

Bowman, Dan. 1968. What makes a lift truck productive? *Plant Eng.* May 16.

Jordan, Wayne. 1965. Simple maintenance checklist reduces lift truck costs. *Plant Eng.* August.

Clark Equipment Co. *Planned maintenance manual.* Battle Creek, Mich.: Industrial Truck Div.

Industrial Battery Div. *Instruction maintenance service manual.* St. Paul, Minn.: Gould National Batteries, Inc.

CD Batteries Div. *General service manual — Motive power batteries.* ELTRA Corp.

Continental Motors. Preventive maintenance — Is it worth the effort? *Maintenance manual.* Elk Grove Village, Ill.: Div. Continental Motors Corp.

Industrial tires data book. Akron, Ohio: The Goodyear Tire & Rubber Co.

LP gas conversion manual. Dallas, Tex.: J & S Carburetor Co.

9 | Layout Planning

Up to now we have been covering just the lift truck. Almost all plants have a lift truck and may be in the process of buying one, therefore the reader will at least know what to look for when he buys one. Since the lift truck is usually part of a total materials handling system and should be planned into the work flow, basic layout planning is covered in this chapter.

Materials handling is a difficult subject to discuss without first talking about plant layout as the layout helps determine many materials handling needs.

In many installations, the product itself is not as important as the methods used in processing or packaging the product. For example, food that is packaged in a can will require different handling techniques from frozen foods that are packaged in plastic bags or boxes. This is not meant to underscore the actual charcteristics of the product such as fragility, size, shape, weight, etc, but rather to emphasize that there are many variables when making a plant layout.

Plant Layout Departments

Unfortunately, responsibility for plant layout is not easy to determine. Often it falls under a central engineering department such as plant, industrial, or manufacturing engineering. Nevertheless, plant layout will often be the responsibility of the plant manager in small plants or perhaps be the responsibility of production supervisors in medium-sized plants.

No matter who controls the layout area, its function is absolutely necessary if a plant is to have an efficient materials handling operation. When making a plant layout, past histories must be incorporated with present knowledge to determine the best materials flow. Therefore, determining equipment efficiency will be only as good as the records kept on the operation.

Since layout helps determine the equipment requirements, then it can be assumed that some engineering function will probably dictate the equipment needs. However, in most plants several departments have some role in the selection of equipment. Although procedures vary from plant to plant, the following personnel are usually involved:

1. Production usually initiates a need or request for equipment.
2. Engineering may also initiate the need through changes in plant layout. The engineering function usually studies the problem and specifies the type of equipment to be procured.
3. Purchasing may obtain the bids and issue the necessary paperwork to obtain the equipment.
4. Management approves the action requested.

In addition, other departments may also influence the selection of equipment:

1. Maintenance: Historical records kept by maintenance may indicate that a product made by one manufacturer is more reliable than a product made by a second manufacturer. Another factor may be that one manufacturer's service facility or maintenance program is better than that of another manufacturer.
2. Production: A machine operator may be required to run the handling equipment as well as operate his machine. Time may be essential, with speed more important than cost savings.
3. Personnel: A machine may be so complicated that few persons can operate it. One particular advantage of a lift truck is its similarity to an automobile. Most workers can drive and, therefore, could operate a lift truck. This statement is not meant to slight a lift truck driving program, as the author feels training is an absolute necessity. Nevertheless, most workers are familiar with cars, whereas most new workers have not been acquainted with lathes, presses, welders, etc.

Lately the trend has been to hire a materials manager. This concept has been around for some time, but with the recent usage of automated machinery in moving materials, a man who has expertise in this area is almost a necessity.

Needless to say, such a man is a rarity and many companies use a committee approach or have a project leader for the purpose of analyzing large materials handling programs. Usually the agenda for such a committee or project leader is as follows:

1. Materials and container analysis: The physical characteristics of the material are studied along with characteristics of the container.
2. Building characteristics: The building itself exerts great influence on a handling project. If ceilings are over 20 ft in height then perhaps a stacker crane may be considered instead of a lift truck. If the height is less than that, perhaps a overhead monorail may be considered. In either case, a lift truck will probably always be necessary for secondary handling. If a lift truck is to be used, then door openings, ramps, floor surfaces, and floor loading will all have an influence on the layout and equipment to be selected.
3. Materials handling activity and cost analysis: Important factors in the handling analysis are the operation, equipment used, distance the material is moved, and the volume of material moved. The cost analysis will include cost of labor and operating costs.
4. Equipment analysis: Several recommendations will come from the study. Labor and operating costs of each alternative as compared to the existing method will be included in the final analysis.

An organizational chart of a project leader's role in making a handling analysis is shown in figure 9-1.

Plant Layout Analysis

A plant layout is made up of the most efficient arrangement of physical facilities and necessary manpower to complete an assigned task. Therefore, layout can be just as important in determining the proper work flow in the office as it can in determining the most efficient manner of manufacturing a product.

A layout department needs a starting point and the best procedure to follow is to make records and layouts of existing facilities and equip-

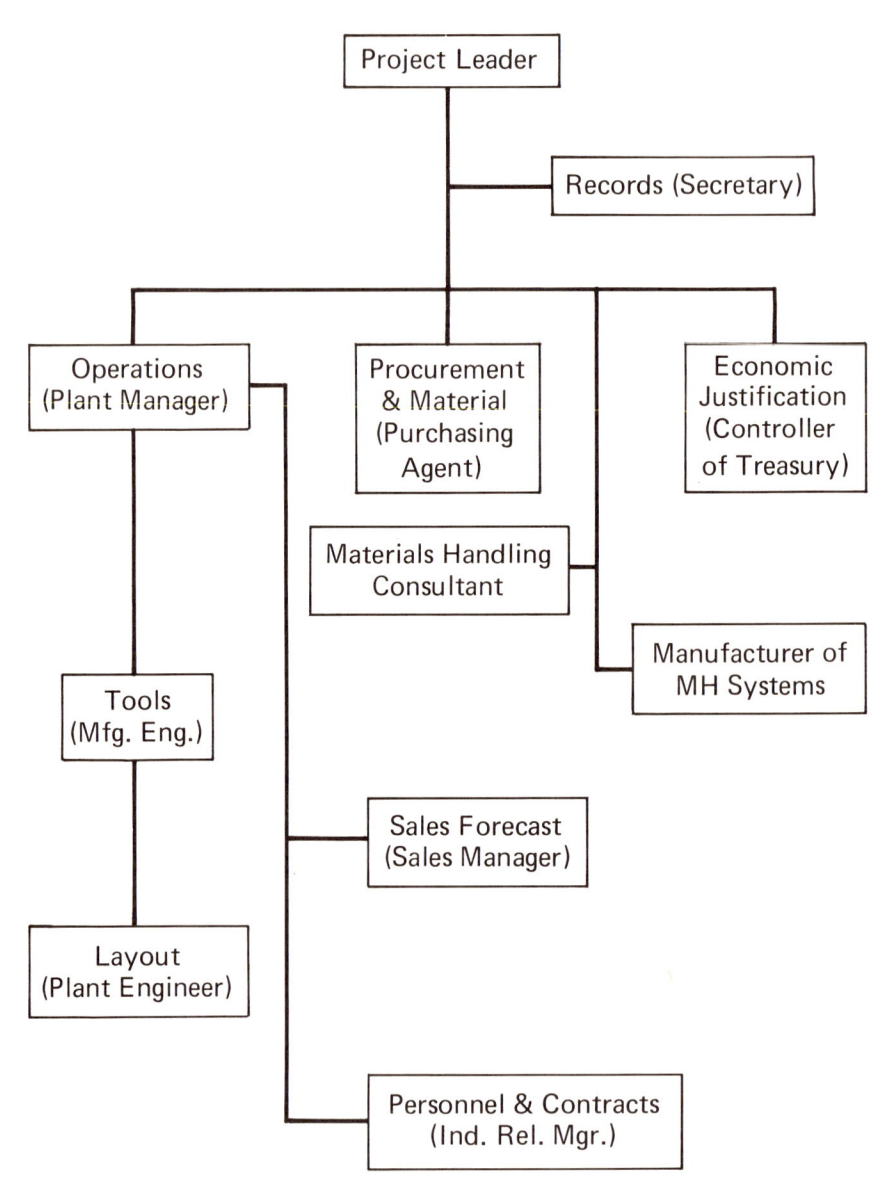

Fig. 9-1. Committee for a materials handling project requires participation from many departments throughout the plant.

ment. Most layouts consist of a floor plan of existing buildings showing the location of equipment. The drawing is usually made on tracing paper or mylar film using a scale of $\frac{1}{4}$ in. = 1 ft. Often templates are used to represent equipment.

Should the plant have a process flow layout or a product flow layout? The process flow routes materials from one machining center to another. A product flow layout is used to route material in a straight line flow through the shop and, instead of using groupings of like machines, all machines necessary to manufacture the one product line are placed in a row (fig. 9-2).

From this starting point, present methods of materials flow can be analyzed. In addition to a floor plan a list should be obtained of all equipment and parts being worked on and moved. From this data, a process flow chart can be generated (fig. 9-3).

Information in a flow chart usually consists of the quantity of parts, the distance the parts are moved, the type of work done, and the type of equipment used. There are five basic symbols used in a flow chart and they are:

Symbol	Shape	Description
Circle	○	signifies an operation. This is a subdivision of a process that involves modifying or changing a part. Action takes place at one specific location.
Arrow	⇨	signifies that there is a change in the location of the part or, in other words, the part has been moved from one operation to another.
Square	□	means an inspection. The part is being checked for quality.
Triangle	△	symbolizes storage. The part remains in an authorized location until called for to be removed.
Big "D"	D	indicates a delay. This is an unproductive event that occurs when conditions do not allow immediate performance or action.

These symbols are accepted almost universally with a few minor exceptions. In the book *Systematic Handling Analysis* by Richard Muther and Knut Haganas, a circle and arrow have been combined to form what the authors call a "handling operation." The authors differentiate a handling operation from a move (or commonly called a transportation) to show the arranging, stacking, positioning, or un-

128 | Lift Trucks

PROCESS LAYOUT

PRODUCT LAYOUT

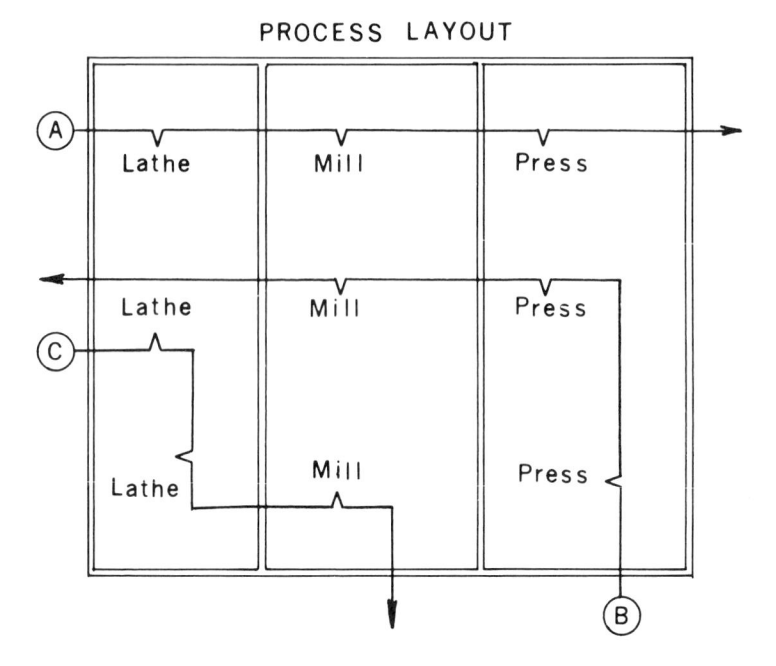

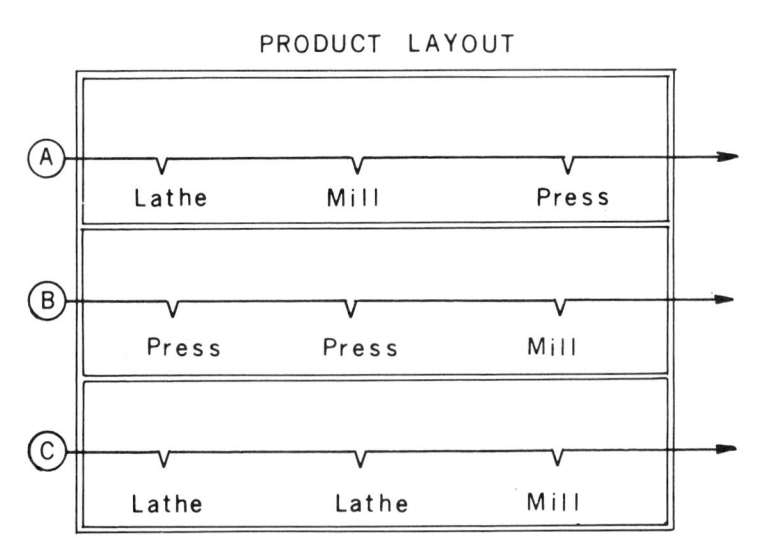

Fig. 9-2. Two different layout plans. The process layout has all like machines in the same area. The product layout is laid out according to the product manufactured.

Step	Operation	Move	Inspect	Delay	Storage	Description of Operation	Dist. Moved	Time Req'd	Equip. Used
1	O	⇨	☐	D	▽				
2	O	⇨	☐	D	▽				
3	O	⇨	☐	D	▽				
4	O	⇨	☐	D	▽				
5	O	⇨	☐	D	▽				
6	O	⇨	☐	D	▽				
7	O	⇨	☐	D	▽				

FIG. 9-3. A process flow chart is the first stepping-stone for making a layout. The product is analyzed according to the various operations performed on it.

loading of materials, which, without this symbol, would have to be placed under an operation or a move.

With the flow chart, present and future layouts can be compared. The flow chart also provides information for making other layout analysis. As an example, information from the flow chart can be transferred to a "From-To" chart. Quite similar to the mileage chart shown on a roadmap, the chart indicates the amount of activity between various departments or functions (fig. 9-4).

The activities can be assigned point values and the location of one department with respect to another can be plotted on a relationship diagram. Various methods are used to make this drawing, but basically it consists of placing the departments that have large amounts of traffic flow between them, as close as possible to one another.

Many layout programs are quite complex. In addition to determining equipment needs, facility problems may also be encountered (such as the layout of utility lines). The author recommends that readers involved in complex programs obtain one of the many textbooks available on plant layout specifically.

Storage Areas

Laying out storage in a warehouse does not necessarily mean that all like materials will be stored in one area, while other like materials are

130 | Lift Trucks

FROM \ TO	rec.	store	cut-off	mill	grind			
rec.								
store	1		4					
cut-off	3	2		3				
mill	4	3	1					
grind	5	4	2	1				

VALUE	CLOSENESS
1	absolutely necessary
2	very important
3	important
4	not very important
5	unimportant

FIG. 9-4. Very similar to the mileage chart on a road map a from-to chart indicates the importance of one operation to another.

stored in another. The four basic considerations in storing materials are as follows:

1. Similarity — This type of arrangement is mainly a repeat of the previous paragraph, that is, all paints are stored in one area, all nuts and bolts in another area, etc. The advantage of such a system is obvious. A man is sent for a load of paint and he goes to the paint section.

2. Popularity — Storing by popularity means just that — the most popular items are stored in one area (closest to the point of usage). For example: In a large warehouse, all fast moving items are located as close to the shipping docks as possible. Since the total loading time of over-the-road trucks will be dependent on how fast loads are obtained, naturally the largest amount of any one specific item that goes on the truck should be located closest to it. If 20 pallet loads of product A are to be loaded on the truck, and only 4 pallet loads of product B, then it would make more sense to require only 20 trips of, say, 100 ft, for A and 4 trips of 300 ft for B.

3. Size — If we are storing 20 varieties of cereal, then package size will have no effect on storage location. However, if transmissions are to be stored with nuts and bolts, then size would greatly affect the lo-

cation of these objects. It naturally follows that if items were stored in a multistory building, small items would be stored on the top floors, while large heavy objects would be stored on the ground floor.
4. Characteristics — Materials can be hazardous, perishable, or of extremely high value. If materials are highly flammable, then these materials should be isolated in a more secure area of the warehouse.

Communications

Worker communication must not be overlooked in layout planning. Whether it be radio receivers on the truck, a PA system, or call-back stations, the worker can be only as efficient as the communications that direct his movement. The advantages of communication are quite obvious. If a truck driver takes a load into storage and is instructed through a two-way radio system, he then can return with a needed item or be directed to another area without first having to return to "home" base.

Records for keeping things in order are an important part of communications. Chaos is bound to occur if the truck driver does not notify the dispatcher as to what he has taken out of storage and where it was placed.

Assigning numbers to storage areas is most important. Even in a small storage area, relying on someone's memory for the location of an item is a poor risk. This will frequently occur in a maintenance storage area and when the part is finally needed, usually the worker who stored it cannot be found.

Probably the best identification system is one with numbers only. This method allows the location of goods to be easily adapted into a computer system. For purposes of identification, one should place himself in the position of the truck driver. When he receives the load what is the first thing he has to know?

First, the driver must know which building and to what floor he should deliver the load (assuming there is more than one building). The next key to a load's place of storage is to what part of the floor the load should be taken. If there are production machines on the floor, then the best identification would be to take the load to a specific bay location. A bay is the area between the center lines of four supporting columns; either the bay itself or the columns can be numbered.

If there are rows of racking, then the next address number can specify a certain row. The row will contain a certain stack number, either on the right or left hand side of the row. Last, but not least, is the particular tier in which the item should be stored.

As a result, a group of numbers can be used to label each and every load stored, without creating confusion for the driver. For example, a group of numbers 062–203–094 could be decoded as: 06, building 6; 2, the floor; 203 could represent either bay or floor; and by using odd numbers for left hand and even numbers for right, 09 would signify 9th stack on the left; and the 4 would signify the particular tier.

Recently, the computer has been adapted for controlling the materials flow in a warehouse. The most complex system exists where the computer is interfaced with the equipment into what becomes the automatic warehouse. While this may seem to be utopia, nevertheless, computers are here and most materials handling engineers or management personnel will want to look into their usage in the handling operation.

There are three levels of control to consider if the computer is to play a role in your warehouse or storage operation. The first area that will become involved is that of management. The computer will be used to schedule and record materials orders, inventories, and shipments.

The next level of control will occur where the computer is used to control the direction of flow through the warehouse. Commands will be issued to various portions (such as a stacker crane) to perform selected functions.

The third level of control will be a completely automatic facility where every operation is under computer control. The sequence of all handling operations will have been programed into the control unit. In addition, maintenance schedules and records will be incorporated into the total program.

Time Studies

Before one can answer the question, "How much does it cost?" one usually has to determine how long it takes to complete the job. Timing an operation is usually the function of the time-study department or

Layout Planning | 133

the industrial engineer. Usually, many standards are available for accurately determining the exact time it takes for a worker to perform any given task (see table 9-1). Each phase of the operation can be timed to a thousandth of a minute. By totaling time values for various motions performed in an operation, the time for a complete operation can be determined without actually going out into the shop and timing

Table 9-1. Basic time values for 4,000-lb electric trucks

	Basic truck motions	Time values (decimal minutes) for various loads in lbs				
		Empty	1,000	2,000	3,000	4,000
Straight runs	Straight					
	1. Forward (Time/ft)	0.0023	0.0024	0.0025	0.0025	0.0027
	2. Reverse (Time/ft)	0.0023	0.0024	0.0025	0.0025	0.0027
	3. Acceleration	0.030	0.025	0.025	0.025	0.025
	4. Stopping	0.020	0.033	0.034	0.035	0.036
	Run in					
	5. 1st level	0.080	0.080	0.080	0.070	0.070
	6. 2nd level	0.080	0.090	0.110	0.100	0.100
	7. 3rd level	0.110	0.120	0.130	0.120	0.120
	Run out					
	8. 1st level	0.060	0.065	0.065	0.060	0.060
	9. 2nd level	0.060	0.065	0.070	0.060	0.060
	10. 3rd level	0.060	0.070	0.070	0.080	0.080
Turns	Right					
	11. Forward	0.055	0.055	0.055	0.055	0.055
	12. Reverse	0.055	0.055	0.055	0.055	0.055
	Right — Stop					
	13. Forward	0.070	0.070	0.070	0.075	0.075
	14. Reverse	0.065	0.085	0.080	0.080	0.080
	Left					
	15. Forward	0.055	0.055	0.055	0.055	0.055
	16. Reverse	0.055	0.055	0.055	0.055	0.055
	Left — Stop					
	17. Forward	0.060	0.060	0.060	0.060	0.060
	18. Reverse	0.065	0.075	0.075	0.065	0.070
Stack	19. Tilt — Bkd. & Fwd.	0.025	0.025	0.025	0.025	0.025
	Hoist					
	20. Up (Time/in.)	0.0028	0.0029	0.0030	0.0032	0.0033
	21. Down (Time/in.)	0.0030	0.0018	0.0018	0.0018	0.0018

134 | Lift Trucks

it. However, many time study men will make their investigation under less than ideal conditions.

Here is an area where layout is most important. In making his run, the driver may encounter other trucks, either in the aisle or approaching his truck from around a corner. Then there is always the possibility of a pedestrian walking in front of his truck, forcing him to make a sudden stop. The end result is that there can be a great discrepancy between the theoretical time and the actual time required to perform the operation.

As an example, the following results were obtained from a time-study test made by a leading truck manufacturer. The test consisted of a truck picking a load from a rack, turning two corners, and placing the load into a truck. Under ideal conditions, trucks from five different manufacturers all made the run in almost the exact same time (the difference between the fastest and slowest was only $1\frac{1}{4}\%$) when no load was involved. However, as soon as a very stable load was added to the truck, the time for the operation increased 9% over the time when no load was carried.

If the load was just stable, the operation took 28% more time than when no load was carried. And if the load was unstable, the operation required 36% more time than under a no-load condition.

When an interference was added to the study, the results were even more startling. With interference (such as an encounter with another truck and meeting a pedestrian) the time to make the run with a very stable load was 33% greater than when the run was made with no load and no interference. When the load was just stable, the time was 56% greater, and when the load was very unstable, it required 66% more time than when no load was on the truck. This held true for almost any make of truck.

As a result, just knowing the time required to perform an operation does not necessarily mean it will be an efficient operation. Good layout and good visibility at cross aisles and proper loading on the truck also are needed.

Actually, an efficient time study man will first analyze the operation and then list any problems the operator encounters. After all, timing a job that is being performed correctly will be a lot easier than explaining a study that was made under less than ideal conditions.

Simulation

With the advent of the computer, layout planning has become more sophisticated. One of the techniques coming into play is the simulation of a materials handling problem.

Actually, simulation is nothing new. In ancient times, models of ships were often constructed before the ship was built. Now with a digital computer simulation, a solution can be done mathematically for materials handling problems.

Unfortunately, simulating a situation is not an easy task. Quite often personnel does not have the talent to design a mathematical model or the company may not have a computer for processing the model.

Nevertheless, the problem can often be solved by using a simulation method called the "Monte Carlo" technique. A typical problem would be analyzing the number of lift trucks required to service a large loading dock area. For example, suppose 100 over-the-road trucks arrived at a large warehouse each day. These trucks arrive during a 10 hour period and each truck requires half an hour to unload. If one were to use averages only, then it would be easy to say that on the average 10 trucks arrive each hour and that unloading one truck requires one-half hour. Therefore, five hours of lift truck time are required each hour or, in other words, five lift trucks are required.

But no one can guarantee 10 trucks are going to arrive each hour. Depending upon the trucks' routings and schedules, five trucks may arrive the first hour, 13 trucks may arrive the next two hours and maybe 19 trucks may arrive just before lunch. Perhaps none of the trucks arrives during the lunch hour, but, again, maybe 25 trucks will come rolling in the hour after lunch. Then we may have five trucks arriving per hour for the next three hours and then, perhaps, 10 trucks during the last hour of the day.

A simple analysis of the problem would be: Since most trucks will arrive just after lunch, we should have about 13 lift trucks (25 trucks per hour times $\frac{1}{2}$ hr unloading time per truck equals $12\frac{1}{2}$, or 13 lift trucks). But, this would be unreasonable — what do you do with 13 lift trucks in that hour when only five trucks need unloading?

136 | Lift Trucks

The preceding example was given not to show you how to simulate, but rather to show the need for simulation.

PERT and Other Management Tools

Often, a plant layout is only the first step in a long chain of events that results in a completed materials handling project or, perhaps, even a new facility. If you were to build a new warehouse facility it would be rather silly to have the lift trucks delivered while the building foundation was being poured. Actually, you may prefer to have some of the lift trucks arrive just as the building is completed so they can be used in transporting new equipment throughout the building. Once the building is completed, then delivery on the rest of the trucks should be expected.

A management tool called PERT is being used to help control the many facets involved in a sizable project. PERT stands for *Program Evaluation Review Technique* and was designed to aid managers in planning and controlling large and difficult programs.

Previously, the flow chart was recommended for planning and laying out a new plant. When using PERT, we develop what might be called "Management's Flow Chart." Its actual name is a *network diagram*.

The chart has three main designs on it: activities, events, and constraints. An activity represents the work effort of a project. It also has direction and has an arrowhead on the line.

An event is defined as the start or completion of a task and is usually represented by a circle. As a result, all activities begin and end with an event.

The constraint indicates the relationship of an event to a succeeding activity. Therefore, in figure 9-5, B cannot proceed until A has been completed. A simple network diagram is shown in figure 9-6.

The previous paragraphs represent a brief introduction to a very useful management tool. Often, a PERT program will be computerized. However, it still takes the human element to organize and program the situation.

In a lift truck analysis, a bar chart can often be used to organize the selection of a lift truck program (fig. 9-7). A bar chart is a simple graph showing the time elements involved in a project. It is a way of organiz-

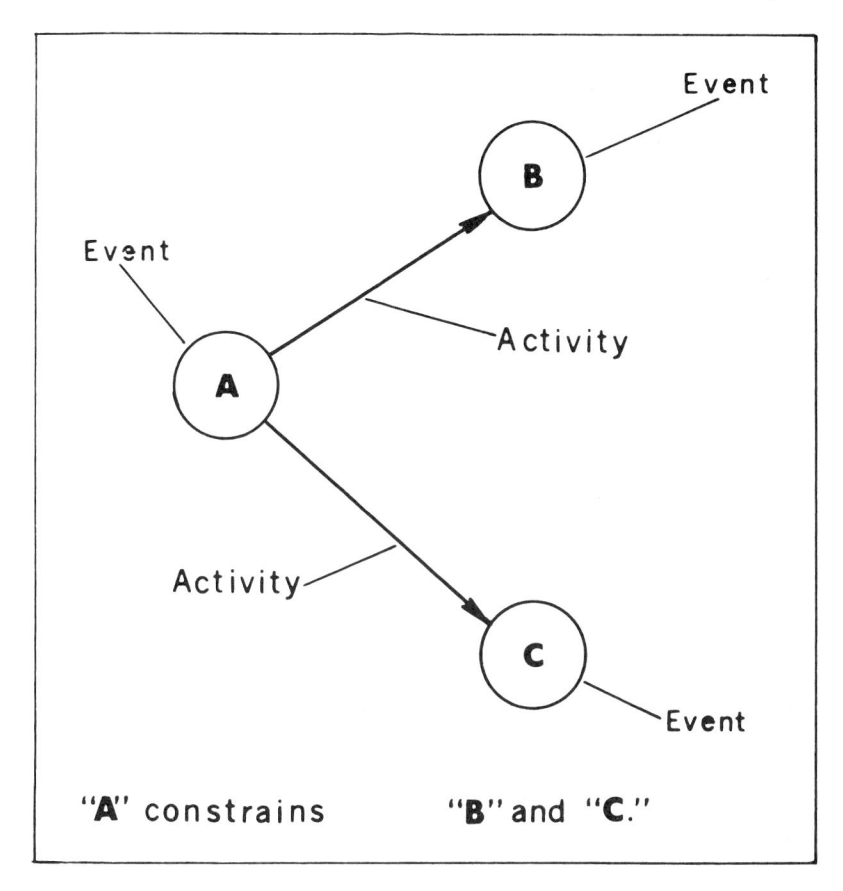

Fig. 9-5. The main elements in a network diagram: an event is the start or completion of a task and is represented by a circle, an activity is the work effort and is represented by a line and given direction with an arrowhead.

ing a task so that steps are not repeated and it also keeps a manager informed as to the progress of the program.

Summary

A plant layout is the first step in organizing any good materials handling program. The layout serves as a plan of action and furnishes guidance in equipment selection. Not only are the equipment needs determined from the layout, but building needs as well.

138 | Lift Trucks

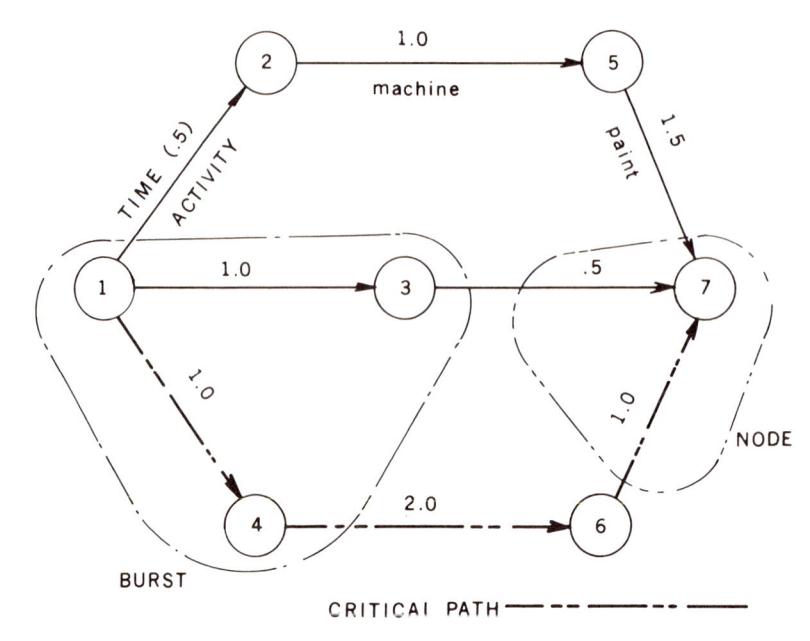

FIG. 9-6. The total network drawing consists of many elements and activities. The work route that requires the most time for completion is called the critical path since the total job cannot be completed until this is done.

Item	Element of Time (Days, Weeks, Months)						
	1	2	3	4	5	6	7
make layout	⌐──┐						
purchase mtls.		⌐────┐					
construct bldg.			⌐────┐				
install utilities					⌐──┐		
paint					⌐──┐		
install machines					⌐─────┐		
train workers						⌐───┐	

KEY ⌐ = Begin   ┐ = Finish   ──── = Elapsed Time

FIG. 9-7. For small projects, the simple bar chart can easily help the manager keep track of his project.

As an example: A new freezer warehouse for a good distribution center may cost upward of $100,000 for a 2,000-sq ft building. By using narrow aisle trucks instead of counter-balanced trucks, only 1,400 sq ft are required, thus saving $25,000 in building costs. As a result, all the equipment can be paid for by the savings realized in construction costs.

Granted, this is an oversimplification of an industrial problem, but the results are valid. Layout planning is necessary for good efficient material flow and any manager who buys equipment without having the problem completely analyzed is only prolonging a bad situation.

Bibliography

Broom, H. N., *Production management.* Homewood, Ill.: Richard D. Irwin, Inc. 1962.

Moore, F. G. and Jablonski, R. 1969. *Production control.* 3d ed. New York, New York: McGraw-Hill.

Briggs, A. J. 1960. *Warehouse operations planning and management.* New York, New York: John Wiley & Sons.

Iannone, Anthony. 1967. *Management program planning and control with PERT, MOST and LOB.* Englewood Cliffs, N.J.: Prentice-Hall, Inc.

Buffa, Elwood. 1969. *Modern production management.* 3d ed. New York, New York: John Wiley and Sons, Inc.

Reed, Ruddell, Jr. 1961. *Plant layout; factors, principles and techniques.* Homewood, Ill.: Richard D. Irwin, Inc.

Muther, Richard. 1968. *Systematic layout planning.* Boston, Mass.: Cahners Books.

Sims, E. Ralph, Jr. 1968. *Planning and managing materials flow.* Boston, Mass.: Cahners Books.

Apple, James M. 1963. *Plant layout and materials handling.* New York, New York: The Ronald Press Co.

Bowman, Dan. 1969. Which best fits your needs—Stackers or lift trucks? *Plant Eng.* April 17.

Mueller, Dale M. 1970. Applying computers to warehousing. *Automation.* Jan.

Sundstrom, J. F. 1969. Simulation, tool for solving materials handling problems. *Automation.* Dec.

10 | Attachments

"It's what's up front that counts," may be true for a cigarette, but it is also true for a lift truck. Previously we have discussed the basics of the lift truck and how to make a materials handling analysis. Now, we get down to the "meat" of the truck — the front end.

There are many variations in fork design from rams used to pick up coil material to chisel forks for slipping under boxes and these devices are usually product oriented. There are also innumerable attachments designed to slip over the forks, the most common device being the fork extension.

Most of the slip-on attachments are specialties and are used in low-volume operations. Among the special attachments are portable booms, drum tip-over devices, and maintenance items, such as snow plows and sweepers. The choice of attachment will be determined by what the truck is to do in the handling system.

Nonpowered Attachments

Power is not required for the operation of all front-end attachments. Simple methods can be used for lifting by merely redesigning the load. For example, in the appliance industry many loads require a very strong container to allow stacking on top of one another. The appliance container usually has sufficient strength to support the weight of its contents. Therefore, if there is a way of locking the truck into the container, lifting can be quite inexpensive (fig. 10-1).

Many cartons are constructed with a folded locking cap. This folded cap becomes part of the side walls, which, in turn, become part of the

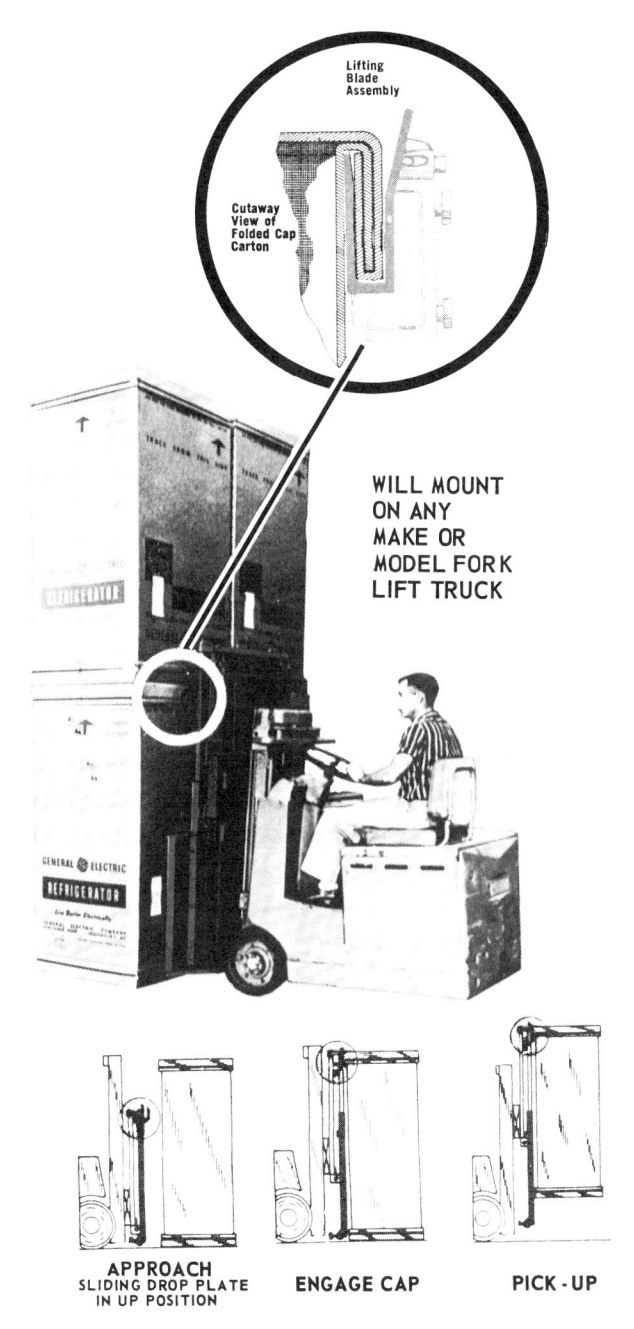

FIG. 10-1. Special carton construction allows lifting at the carton top. This eliminates the need for a clamp or pallet. Courtesy of Basiloid Products Corporation.

141

carton bottom and this construction permits mating of carton and truck. By inserting a lifting blade into the cap top, the whole load can be lifted without the use of forks.

Powered Attachments

There are many types of front-end attachments. In fact, about their only limitation is the ingenuity of the engineer in specifying his needs. The sideshifter, of all the powered attachments, is the one that is most widely used (fig. 10-2). Sideshifters can be used in almost any loading,

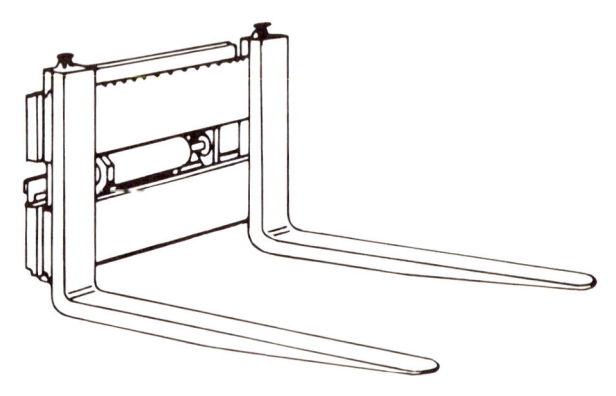

Fig. 10-2. Sideshifting attachment allows load to be shifted laterally, thus making truck maneuvering easier.

unloading, or warehousing operation. Their chief advantage is that they allow a lateral (sideways) movement of the load without moving the truck.

Lateral movement permits a truck operator to greatly reduce his maneuvering time when placing loads in tight positions. Loads can be stacked close together and in trucks or boxcars, and loads can be placed against the car sides without scraping the walls. Hydraulically operated, most sideshifters allow at least 6 in. of lateral movement. In large trucks some units can be obtained with up to 12 in. of movement. Any additional motion could create an unbalance between the center of the load and the center of the truck. A competent driver should be able to place a load within 6 in. of its required position.

Attachments | 143

If the plant is large and there are several lift trucks in the fleet, at least one truck should be provided with a sideshifter. Often, the maintenance department is called upon to move machinery and a sideshifter will greatly facilitate positioning the machine. A sideshifter can be used with most any other front-end attachment. Always remember its main function – to provide a lateral movement of a load – not as a device to open doors or nudge equipment.

The next attachment most commonly used is a rotator or inverter (fig. 10-3). This device allows the forks or other attachments to rotate the load. Rotation devices are available with 180 degrees' rotation or a full 360 degrees' continuous rotation in either direction.

When used with standard forks, dumping, inverting, or positioning operations can be performed as long as the container will remain on the forks. If the container does not have sleeves for the forks to slip into, as for example in the case of a 55 gal. drum, then the forks will have to be replaced with clamp, vacuum, or other special attachments to securely hold the load (fig. 10-4).

There have been many advancements in attachment design over the

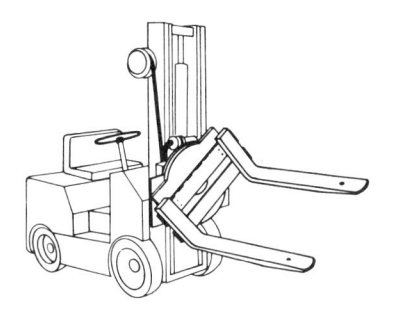

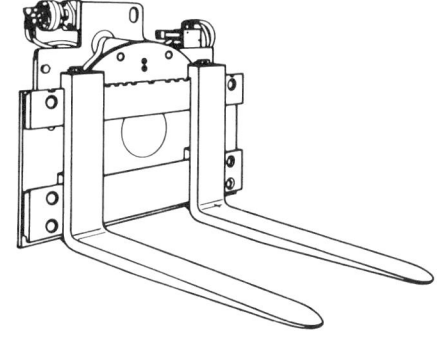

FIG. 10-3. Rotator allows inverting of loads. Container requires "fork-holds" to prevent container from falling when tipped over.

144 | Lift Trucks

FIG. 10-4. Rotator with clamp allows a truck to position loads. Courtesy of Allis-Chalmers

past years. In high-volume operations, the cost of the attachment is recovered in a relatively short time through pallet, space, or time savings.

The two major power attachment operating principles are clamping and vacuum action. Clamping action can be used on any type of product or carton that has enough strength to counteract the clamping force. If the surface area is large in relation to the product weight, as in the case of cartons of tissue paper, the actual pressure applied in psi can be quite small. Often the pressure exerted by the clamp on the carton is less than that received in stacking — only about 2 psi (fig. 10-5).

Almost all case goods, with the exception of weak-walled containers (such as bagged flour and sugar), can be clamped. Clamping an item is slightly slower than picking up a load on a pallet, but the truck will be faster in overall travel time due to the good load stability provided by the clamping action. It should be noted when handling cases, a clamp truck need not pick up the whole load. However, the bottom tier of a load should be properly stacked and it is advisable to have each tier on top complete.

FIG. 10-5. With carton clamps, lifting pressure is distributed over a large area, thus actual clamping pressure is not damaging to the product. Courtesy of Clark Equip. Co.

All devices do not clamp directly onto the product. One, usually referred to as a "push-pull" device, clamps onto a craft board sheet placed under a load, then retracts and pulls the load onto the forks. A pushing device is also incorporated into the attachment and the load is unloaded by pushing it off the forks. Other clamping devices can handle drums, textile bales, and paper rolls.

Although not used extensively, vacuum attachments merit some discussion (fig. 10-6). Lifting by vacuum is quite simple. Using the suction cup as an example, by decreasing the air pressure on the suction cup side of an object, the higher pressure on the opposite side will force the object against the cup and hold it in place. The object will be held in place until the pressure is equalized again. This pressure will be equalized if the cup loses its seal or if the material is so porous that it cannot maintain a pressure differential.

Many materials can be lifted by vacuum, including newsprint, craft

146 | Lift Trucks

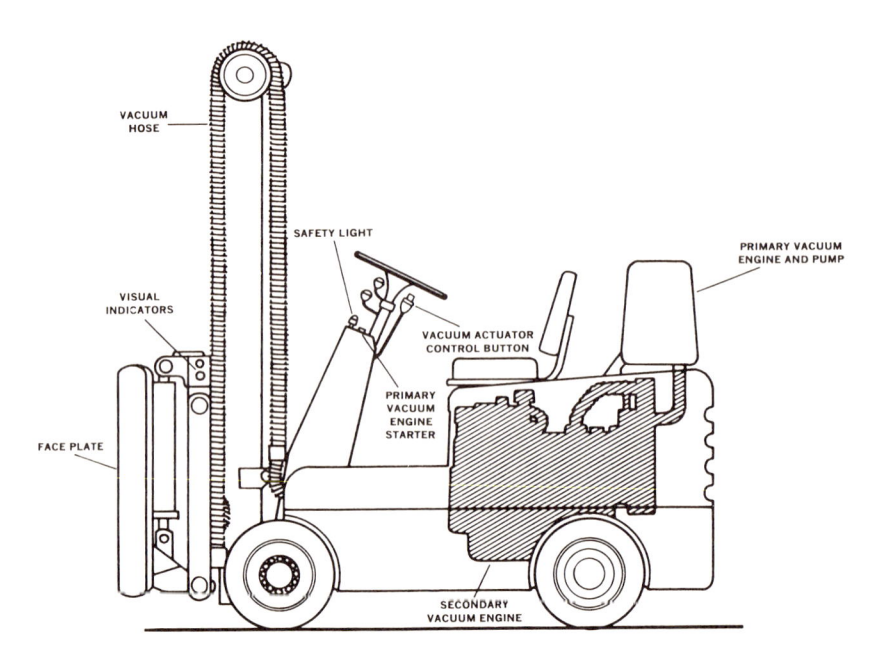

FIG. 10-6. Many vacuum attachments require a power source separate from the truck engine; but in some cases vacuum is drawn off the truck's engine manifold.

liner board, and other roll products. In addition, many items that have a nonporous package, such as appliances, furniture, and steel and fiber drums, can be lifted by surface vacuum.

While vacuum handling is not the solution to all materials handling problems, some of vacuum's important features are:

1. Load stability: The pressure around the object, while it is against the vacuum faceplate, creates a single object.
2. Load protection: Because only a soft rubber plate is in contact with the load, there is no danger of metal-to-product contact as in the case of forks or clamp arms.
3. Space savings: Space required for pallets, clamp arms, forks, etc., is eliminated by vacuum handling. The only object in front of the truck is the load, which allows more maneuverability.

In summary, attachments can be designed to duplicate almost any conceivable motion, whether horizontal, vertical, rotating, or some spe-

Attachments | 147

cialized function (fig. 10-7). They can be used on gasoline, electric, and diesel powered trucks. With the exception of vacuum units, an attachment does not impose any abnormal demand on standard electric truck batteries.

In any attachment analysis, the load itself must be checked first. What is the fastest way to move the load safely and how can it be incorporated into the work flow? What is to be done with the load? Should it be carried, rotated, pushed, or stacked? In turn, one often can find an attachment that will take care of this load (fig. 10-8).

Fig. 10-7. Special attachment allows lift truck to pick up a load that would usually be handled by a sideloader truck. Courtesy of Towmotor Corp.

148 | Lift Trucks

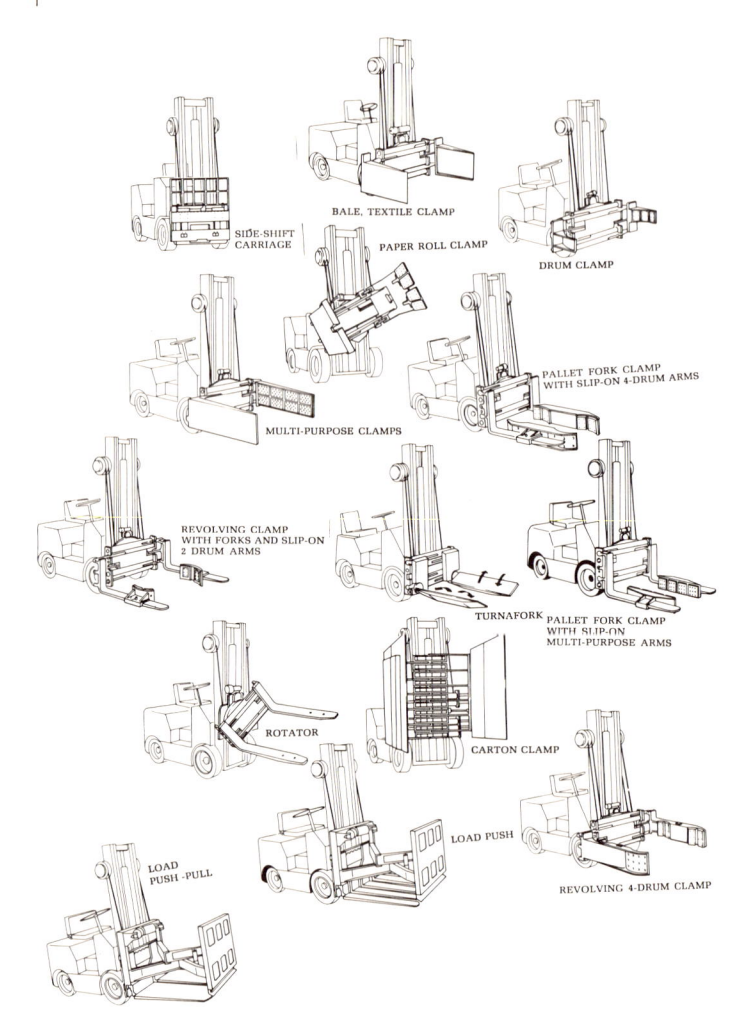

Fig. 10-8. The various attachments available are summarized in this illustration. Courtesy of Cascade Corporation.

11 | The Unit Load, Pallets and Packaging

Hand in hand with attachments is the packaging method of the goods being handled. Some packaging methods will actually dictate the attachment selected. Included in this chapter is the packaging factor, which takes in pallet design and an explanation of the "unit load."

The unit load takes many factors into consideration. Related to the unit load is your selection of a front-end attachment for your truck. Needless to say, a unit load cannot be determined without first analyzing the product itself.

There are various definitions for a unit load. Some persons in industry feel that in order for a load to be considered a unit load, the load must rest upon some support, such as a pallet. However, most persons think of unit loads as all loads which are developed by the application of the principles of standardization and resulting in an optimum load design for the combined activities affected, whether it rests upon a pallet or not.

Nevertheless, a load must be convenient to the user. Therefore, to the housewife, a can of peas may be a load. To the stockboy in a grocery store, a case of peas is a load. In industry, mechanical aids are used and loads can be based on the size of the equipment used in the shop. Therefore, if a lift truck has a 4,000-lb capacity, all loads should be as close to 4,000 lbs as possible.

Naturally, there are limitations to the size of the load. Different materials have different densities. Therefore, a 4,000-pound load of steel will have far less volume than a 4,000-pound load of lumber. As a result, unit loads must be placed in a perspective of weight to volume ratio. Added to this ratio is the shape of the load. A gasoline engine presents far more problems in handling than 100 boxes which would take up the same space. If the load is liquid or loose (fig. 11-1), it will have to be contained. Even if the load is contained, it may be desirable to stack the load (fig. 11-2) or even place it in a larger container.

Goods placed on a pallet or a skid probably constituted the first widely accepted method of making a unit load. With this method as a starting point, we can follow the development of the unit-load concept.

A skid or pallet provides a fixed area (length and width) for a specific amount of goods. The load may be in drums, cartons, bags, or other types of containers. As long as the load is to have horizontal movement only, either a pallet or a skid can be used. However, if it is necessary or desirable to stack the loads, then a pallet will be used. A skid platform has a deck surface and two runners, while a pallet has a deck and a bottom surface that is stackable. If a skid is to be stacked, it has to be placed in a storage rack. In general, a platform truck is used on skids and a fork truck is used on pallets. However, the fork

Fig. 11-1. A singular load is easily placed on a pallet. Courtesy of Clark Equip. Co.

Fig. 11-2. A unit load can be easily placed on a pallet. Courtesy of Allis-Chalmers

can be used to lift a skid. Figure 11-3 shows the difference between skids and pallets. If the load is stackable, a fixed height can be determined. As an example, four drums may be placed on a pallet providing a unit load. This procedure provides a surface whereby another pallet load of drums may be placed on top of the first load.

Some items, such as cartons, are stackable according to the limits of the strength of the carton material and the stackability of the load. If the load is in powder or liquid form, instead of in drums, sidewalls can

152 | Lift Trucks

a.

WOODEN SKID
(with or without metal frame)

b.

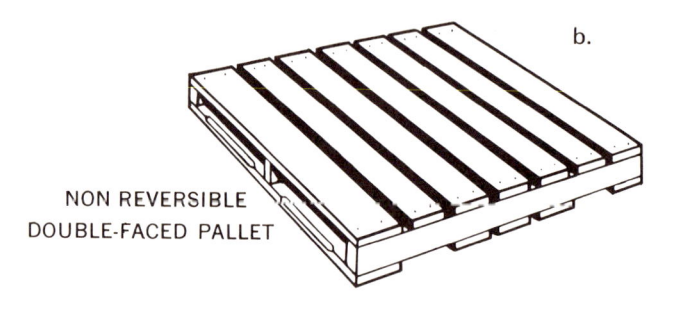

NON REVERSIBLE
DOUBLE-FACED PALLET

Fig. 11-3. A skid (*a*) has a deck surface and runners, while a pallet (*b*) has a top and bottom surface separated by stringers.

be attached to the pallet or a tub can be placed on it to contain the load.

There are two basic types of pallets: two-way entry and four-way entry. The basic two-way pallet consists of a top deck and bottom boards with three stringers. The outer two stringers are flush with the edge of the deck and the third stringer is centered. There are several variations, but, basically, a two-way pallet allows entry from two sides only (fig. 11-4*a*).

Four-way pallets differ from the two-way pallets in that they allow a lift truck to pick up the load from any side (fig. 11-4*b*). As in two-way pallets, there are several varations in the four-way pallet category (fig. 11-5).

One other type of pallet that may be encountered is the slave pallet. This pallet is usually a specially designed board used in an automatic storage and retrieval system. Since it was designed for just this particu-

The Unit Load, Pallets and Packaging | 153

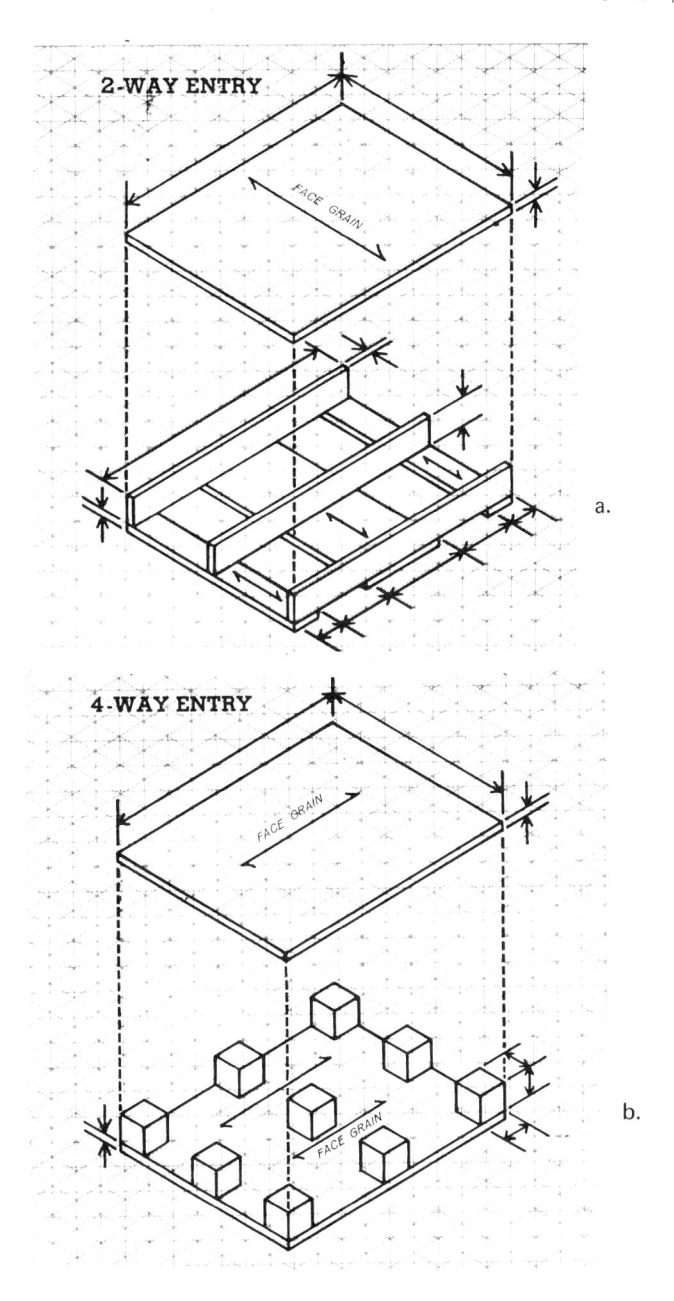

Fig. 11-4. A two-way pallet (*a*) allows entry from only two directions, while a four-way pallet (*b*) allows entry from any side.

Double face
non-reversible
pallet

Double face
fully reversible
pallet

Double wing
pallet

4-way Notched
stringer pallet

Single wing
pallet

4-way Block
type pallet

Expandable
pallet

Pallet stacking

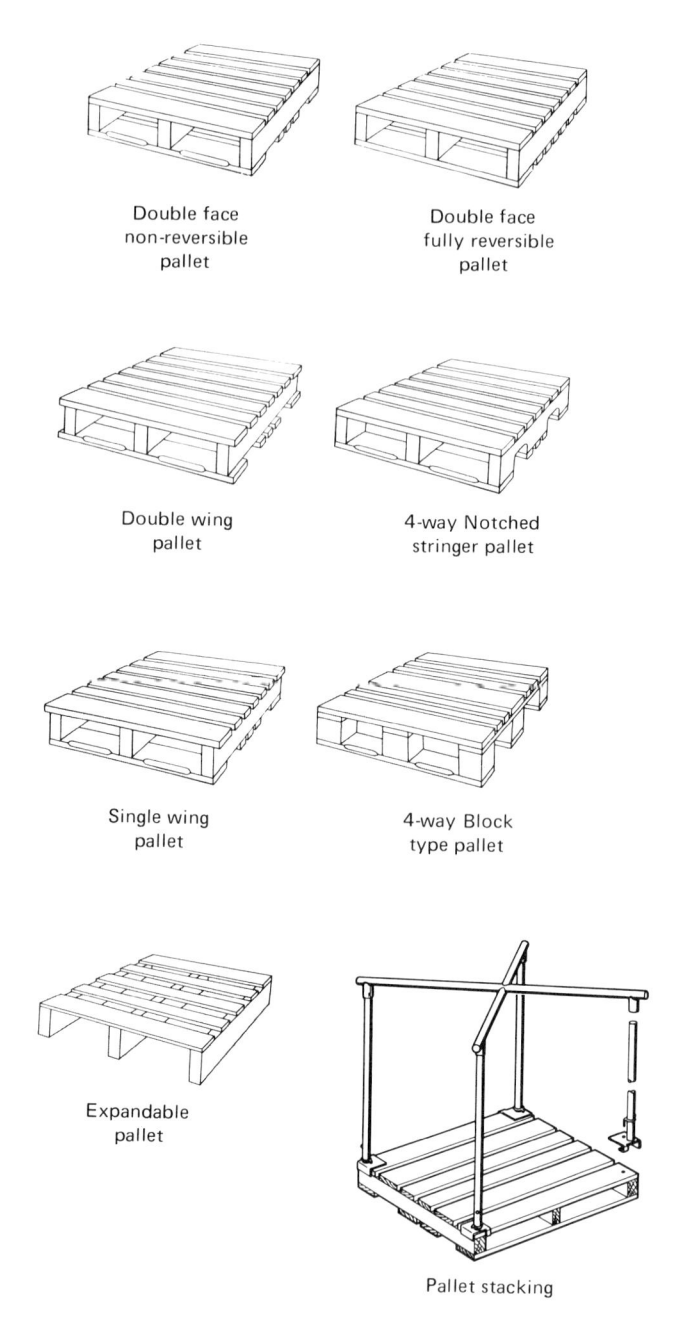

Fig. 11-5. Pallets come in many forms, several of which are illustrated. Shown in the last example above is a pallet stacking frame, which allows the stacking of broken-case loads.

lar system, it is a slave (or captive) to the system. In some cases a fork truck may be used to transfer goods from this type of system. Unfortunately, when a pallet serves a dual role (that is, to an automatic system and for general purpose movement) it often becomes misplaced.

In analyzing the load, another consideration is its final destination. Will the load be placed in storage or will it be worked on within a short period of time? As in the case of castings in a machine shop, it may be best to carry the material in a drop-bottom dump box (fig. 11-6*f*). This allows the load to be easily dumped at a work station and the container can be used elsewhere.

There are many different styles of containers based on the pallet or skid principle. Pallet boxes can be made out of wire, wood, corrugated board, fiber board, sheet metal, and various combinations of materials. They may be collapsible, allowing many to be stored in the place of a few when not in use. Other containers have special openings or dumping mechanisms, allowing contents to be quickly unloaded. Various pallet configurations are shown in fig. 11-6.

When the handling of unit loads involves the use of pallets, the following factors should be investigated when analyzing the different types available:

1. Deck: Is a solid deck required? Does it have to be smooth to prevent bags or cartons from tearing?
2. Durability: Will the pallet receive extremely rough handling or travel through areas that are wet, cold, corrosive, or hot?
3. Weight and space: The weight of a pallet has no great effect on the lifting capacity of a lift truck. However pallet weight may add considerable cost to a truckload or railcar load of product. Likewise, the pallet height has no effect on the truck, but too high a pallet may cause problems where goods are stacked in a low-ceiling area.
4. Turnover and standardization: If the pallets remain in your plant, the size can be tailored to fit your needs. If the pallets are shipped out, then they should be compatible with the shipper's carrier or your customer's needs. If the pallets are nonreturnable, then cost should be held to a minimum.

Important as the pallet is, the method of placing the load on the pallet is just as important. Lift truck travel and handling time will increase if material is stacked in an unstable manner. To prove this

156 | Lift Trucks

Steel Skid Box

Gas Cylinder Pallet

Wire Pallet Box

Wooden Pallet Box

Collapsible Pallet Box

Dumping Hopper Pallet

Bulk Container Pallet

Stacking Pallet Frames

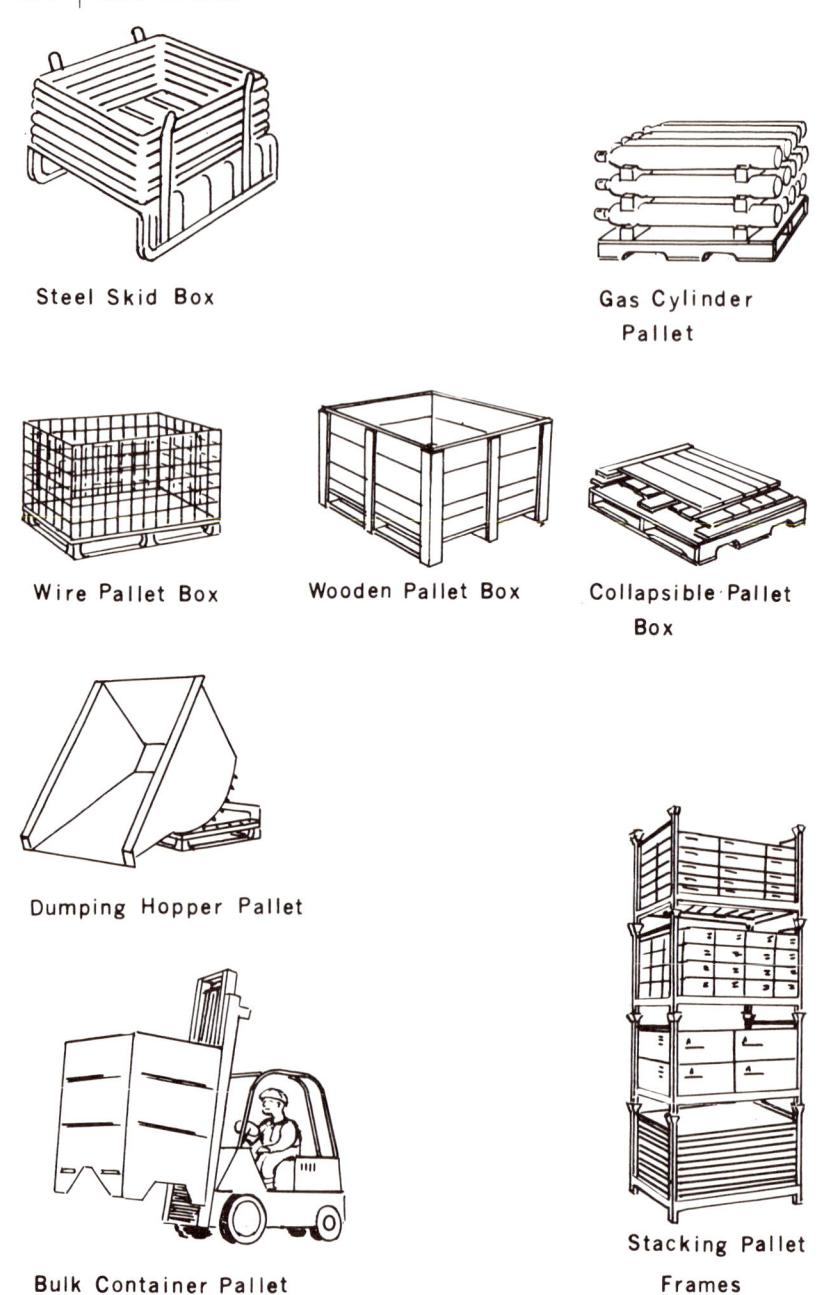

FIG. 11-6. Many pallet and skid configurations are available. Not shown are wirebound or corruguated pallets.

point, a lift truck manufacturing company tested load stability. During the test, travel time increased over a base time set by an unloaded truck over a specified distance by the following amounts: When a very stable load was added to the truck, travel time increased 9%. If the load was just barely stable, the time increased 28%, and if the load was unstable, travel time increased as much as 36%.

Stability can be increased by the stacking pattern used when setting goods on the pallet. Banding the topmost layer of the load also increases stability. Recently, shrink-film packaging, which offers the utmost in stability, has been applied to pallet loads. After the pallet is loaded, a plastic film is placed over the load and is drawn tight in a special oven (fig. 11-7). The film not only stabilizes the load, but protects it against any moisture damage as well. Stacking methods are shown in Figure 11-8.

As mentioned earlier, paper packaging materials tend to lose their strength when stored over a period of time. If this is the case, then pallet racks or pallet stacking frames may have to be used. There are many styles of pallet stacking frames, but they all serve the same function. They are special members which fasten to a pallet to enable the stacking of nonuniform or nonrigid materials.

Packaging Materials

A package is a method of unitizing a specific amount of a commodity. The housewife buys a can or two of peas, a size convenient for her. Placing many cans in a case makes it convenient to the supermarket. Many cases placed on a pallet makes shipping and warehouse storage more convenient.

The supplier of peas to the food processor may deliver the material in large containers or even in bulk car lots. Therefore, goods, ideally, are packaged according to the convenience required in processing and handling.

The package also provides a method of identifying the load. Again referring to the housewife, she can see a picture of peas on the label and that is usually all she needs. If she needs a specific amount, she can check the size and volume on the label. The supermarket requires that the carton the cans come in be identified and will want the quantity and date of manufacture listed. Any special warnings, such as for perish-

158 | Lift Trucks

Fig. 11-7. The latest in load stability: a plastic film is placed over load and the unit is sent through an oven which causes film to "shrink" tightly around load. Product temperature is raised only a few degrees and process does not harm paper or other combustible materials. Courtesy of Poly Plastic & Design Corp.

able products, should also be printed on the carton or even on the can, if necessary.

If the case comes from an automated warehouse, reflective coding tapes may be attached to the carton. Identification is a necessary part

The Unit Load, Pallets and Packaging | 159

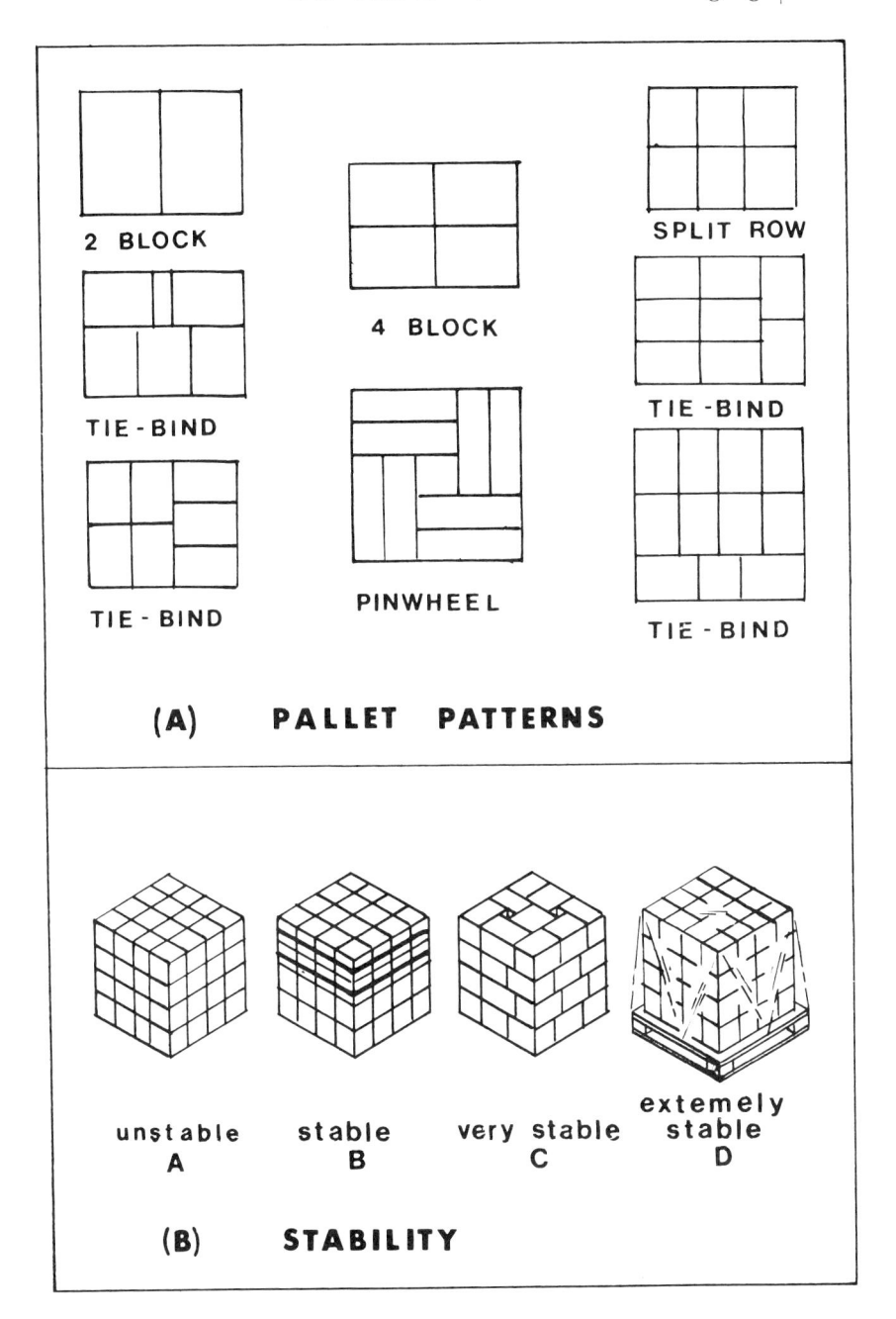

(A) PALLET PATTERNS

2 BLOCK

TIE-BIND

TIE-BIND

4 BLOCK

PINWHEEL

SPLIT ROW

TIE-BIND

TIE-BIND

(B) STABILITY

unstable
A

stable
B

very stable
C

extemely stable
D

Fig. 11-8. Load patterns are shown in (*a*) and stability in (*b*).

of packaging and the type depends on the manufacturing processes and how the product is eventually sold. As a result, all identification should be well protected and properly located for it to be read by a person or decoded by an automatic device (fig. 11-9).

Packaging provides a means to display and sell goods, as well as to protect the contents. Extreme care should be taken to maintain the package's appearance whenever it serves as a display case for the product.

When material is stored in paper or corrugated cartons, temperature and humidity determine how high and for how long the goods may be stored. Paper materials will experience fatigue or will creep, if stored for any length of time. Eventually, this can cause the cartons or bags to fail, which, in turn, can bring about the collapse of an entire stacked pile. The improper amount of moisture in the air is more detrimental to stacking height than the temperature and therefore must be controlled (fig. 11-10).

Paper Shipping Sacks. Shipping sacks come in a wide variety of shapes and sizes (fig. 11-11). They are constructed of from one up to six layers (walls) of different papers or other materials. Treatment of the paper depends upon whether the sack is to prohibit the contents from absorbing moisture; prevent chemical action from taking place with the contents; protect the contents from insects and vermin; or prevent loss of moisture.

Empty sacks should be stored in a fairly humid atmosphere so that they retain their flexibility and strength. Average temperature should be 70° F and relative humidity should be 50 to 60 percent. Sacks should never be stored where there is extreme heat or cold; either extreme will rob paper of its moisture. The sacks that have been on hand the longest period of time should be used first in the filling operation.

Sacks are usually placed on a pallet when they are moved by a lift truck. Sacks should be interlocked by alternate stacking and if an overhang is unavoidable, it should not be more than 2 inches. The pallet deck should be as smooth as possible to avoid tearing the sack. Often, special adhesives or tapes will be used to help hold the load together. The glue on the sack will prevent slippage yet allow the sack to be easily lifted without sticking to the one beneath it.

Many other materials, including cloth and plastics, are used in sack construction. Although sacks are usually of a size that allows a man to

The Unit Load, Pallets and Packaging | 161

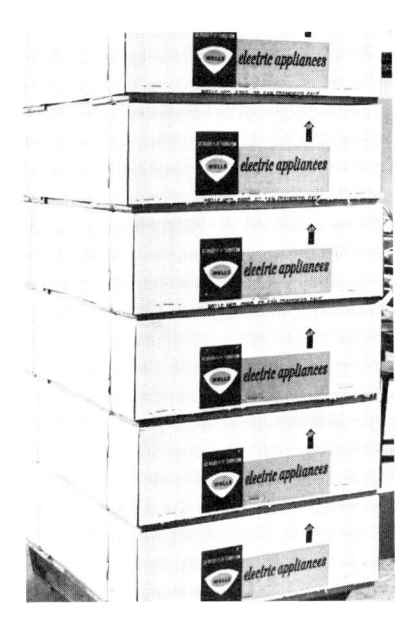

FIG. 11-9. Packaging must identify as well as protect. Courtesy of FMC Corp.

162 | Lift Trucks

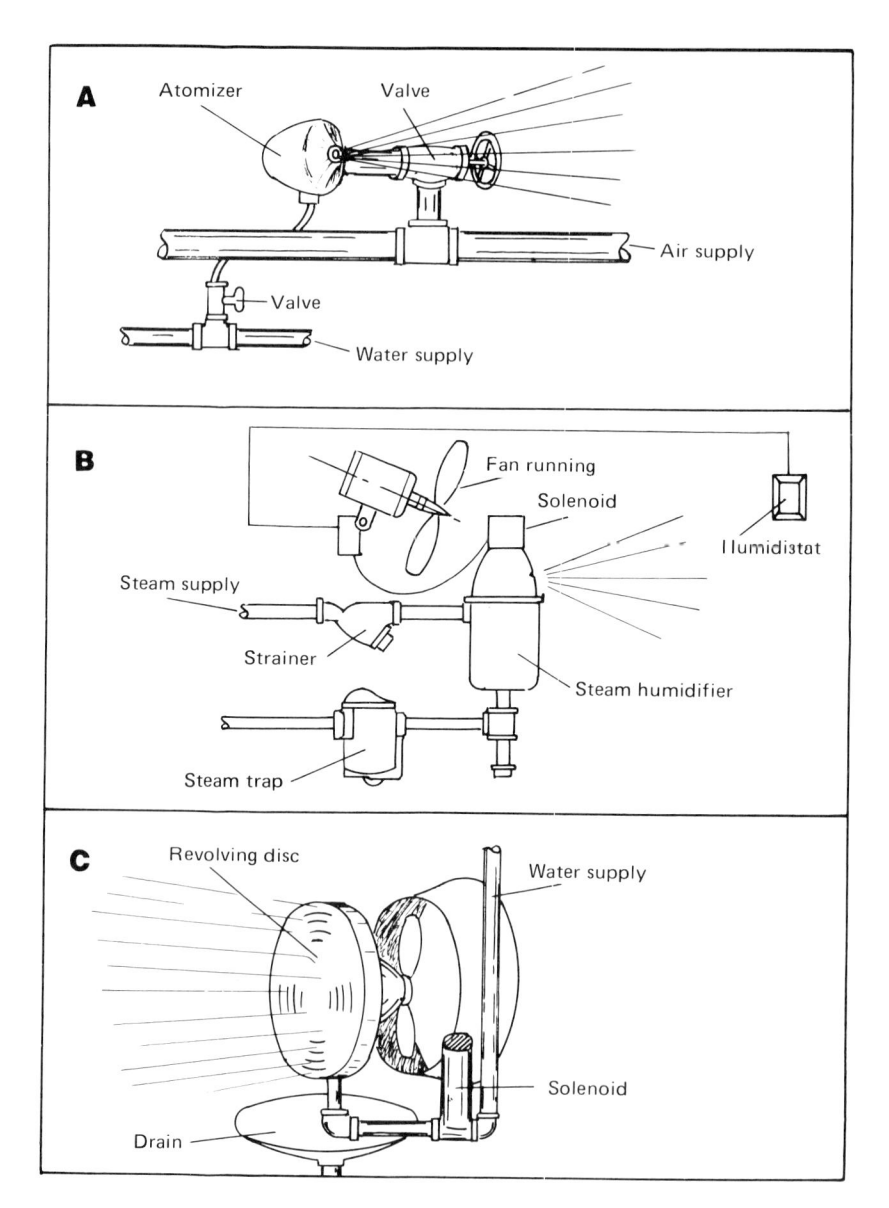

Fig. 11-10. Three methods of adding humidity to a room where paper products or corrugated board are stored: (a) Water atomized by air jet; (b) steam; (c) water atomized by a mechanical impeller.

The Unit Load, Pallets and Packaging | 163

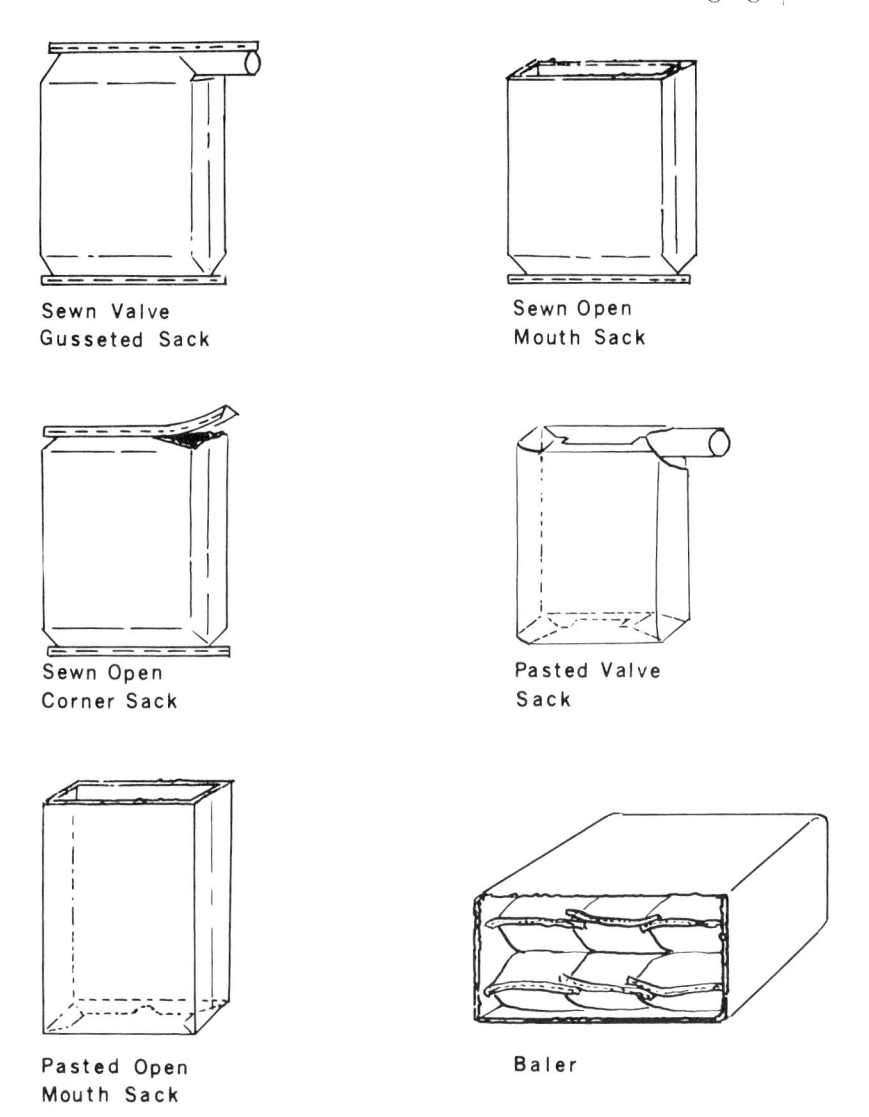

Sewn Valve
Gusseted Sack

Sewn Open
Mouth Sack

Sewn Open
Corner Sack

Pasted Valve
Sack

Pasted Open
Mouth Sack

Baler

FIG. 11-11. These are just a small sampling of the many paper and plastic sacks available.

lift one unit, there have been instances where sacks have been designed to hold several thousand pounds.

Corrugated Boards. One of the more popular types of shipping containers is the corrugated box. Construction of corrugated boards con-

164 | Lift Trucks

sists of a fluted sheet glued to one or more liners. Structural characteristics are governed by the liner, corrugated medium, height and number of flutes, and number of walls. Experiments have been made using steel and aluminum foil as a fluted medium, to provide greater strength; however, normally the box will be entirely of paper products (fig. 11-12 and table 11.1).

Various coatings have been developed over the years to give the board more desirable properties. Waxes and hot-melt coatings allow a rigid-when-wet construction. Since many cartons are eventually used as a display center, extreme care will be required in handling this particular type of carton.

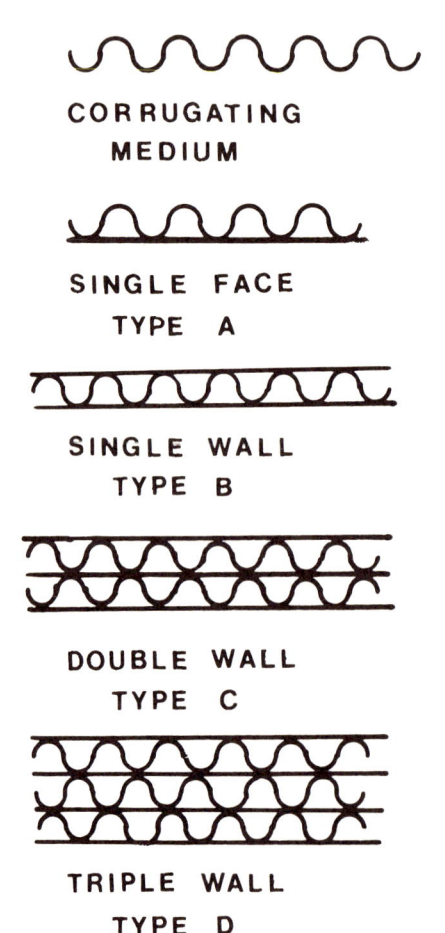

CORRUGATING MEDIUM

SINGLE FACE TYPE A

SINGLE WALL TYPE B

DOUBLE WALL TYPE C

TRIPLE WALL TYPE D

FIG. 11-12. The flute characteristics of the corrugated board determine the carton strength. See also table 11.1.

Table 11.1. Flute characteristics (See also figure 11-12.)

Type	Flute/ft	Flute ht (in.)	Board thickness (in.)
A	35–37	0.185	3/16 to 7/32
B	50–52	0.105	1/8
C	41–45	0.145	5/32
D	90–96	0.085	5/64

The patterns of stacking can also greatly affect the stability of a unit load of filled cartons. Often, the topmost layer will be capped or strapped.

Wood Boxes and Crates. Wood is used in the construction of shipping containers in many ways from corner supports to complete boxes. Wood is used when the utmost protection is required for a load. Extensive use of wood is required in shipment of machines and in overseas shipments.

Just as there are many varieties of corrugated boxes, there are also many varieties of wood crating materials. Wood may provide the basic frame for a crate with corrugated board or plastic film used as a protective covering. Or, the crate may be constructed of a box grade plywood with a paper coating. Since there are so many variations in crating, it is almost impossible to classify them.

If wood is used in large amounts, it is advisable to have a specialist design the container (fig. 11-13). Many companies operating their own box shops use heavier grades of wood than are necessary, thus incurring excess shipping costs. Often, they also neglect to take advantage of modern assembly techniques. Automatic staplers and nailers can cut costs tremendously. For example, in one plant, one man found he was able to do the work of six workmen simply by using a mechanical nailer.

Drums and Cylinders. Although the standard steel drum will be around for a long time, many manufacturers are taking a second look at this type of shipping container. Corrugated boxes with plastic liners as well as fiber drums are providing newer means for shipment (fig. 11-14). Many other large containers also provide an economical shipping package. The big advantage of this type container is that it can be fitted with a hopper bottom or special opening to tie it into the product process.

166 | Lift Trucks

Fig. 11-13. "Knock-down" wood container that fits on a pallet allows easy return shipment of container. Courtesy of Contain-A-Pallet.

Cylinders often contain explosive gases or corrosive chemicals and require careful handling. Great care should be used when handling cylinders by a lift truck. Special pallets or skids should be used and protection should be given to the valve or cap end of the cylinder. If the valve end is accidentally knocked off a cylinder of compressed gas, the unit will take off like a rocket. If a cylinder is carried in an upright position, it should be chained to a stable vertical member of the truck or pallet to prevent tipping.

Up to now, we have assumed that the loads will be placed on a pallet

The Unit Load, Pallets and Packaging | 167

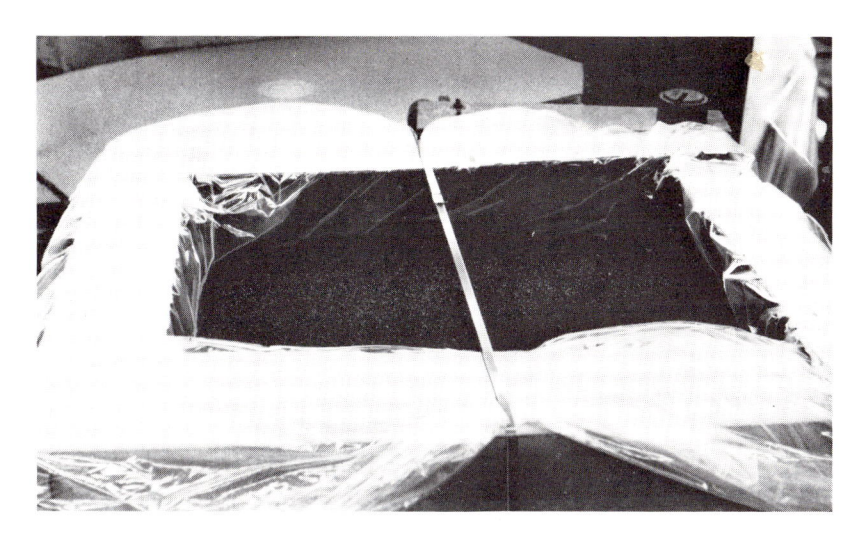

Fig. 11-14. Corrugated boxes and fiber drums with plastic liners often provide satisfactory means of shipment.

168 | Lift Trucks

or skid. However, a pallet adds height and weight to a load. As a result, some companies have turned to other techniques for transporting unit loads (fig. 11-15). Many methods are available and most are related to the front-end attachment on the truck. The following method is mentioned as it shows the economics involved in making a packaging study.

A case history can best illustrate the importance of coordinating the proper packaging method with the right handling technique. Clamp handling of packaged goods got its first big start in 1955 by the Procter and Gamble Co. In their search to reduce warehousing costs they made an extensive study of this subject. Their study resulted in the following conclusions:

Fig. 11-15. Clamp handling is often a good technique to use with packaged goods.

The Unit Load, Pallets and Packaging | 169

1. Wood pallets are still essential for items usually handled in less than unit-load quantities. They are used especially for low turnover items, bags of flour, potatoes, and charcoal; selection and transfer of outbound store orders; and handling store supplies.
2. Clamp trucks are used for high turnover case goods as they have the lowest operating expense.
3. Careful consideration must be given to the end receiver of the goods. He may not have the equipment to unload palletless unit loads.

A cost comparison of pallet and clamp handling is shown in Figure 11-16.

Trends in Containers

In this chapter we have mentioned the packages, pallets, and containers presently in use in the average industrial plant. In the future, packages and containers will tend to grow in size (fig. 11-17). Containerization is now used quite extensively in shipping. Rail carriers and truck carriers are starting to use containerization and, with the advent of the jumbo jets, its usage will increase.

As far as industry is concerned, it all means that bigger lift trucks will be needed to handle the larger loads. The side-loading and straddle carrier trucks will become competitive with the front-end lift trucks — and in competition with them all will be movable yard cranes and overhead cranes.

170 | Lift Trucks

INITIAL INVESTMENT COMPARISON*		
FORK LIFT WITH WOOD PALLET SYSTEM		
Fork Lift Truck	$6,000	
plus		
Wood Pallets (3,000 @ $2.70*)	8,100	
		$14,100
CLAMP SYSTEM		
Clamp Truck Attachment (Conversion of existing equipment)		$ 3,500
or		or
New Clamp Truck		$10,000

*This is a minimum cost.

HANDLING SYSTEM OPERATING EXPENSE COMPARISON*
(10,000 cases per day, or 333 unit loads)

	Wood Pallet	Clamp
A. WAGES PER DAY		
1. Deposit pallets on dock	$ 4.60	None
2. Handle unit loads in warehouse	$ 33.00	$34.00
3. Handle pallets when empty	$ 5.90	None
4. Wage overhead at 20%	$ 8.70	$ 6.80
Subtotal	$ 52.20	$40.80
B. PALLET EXPENSE PER DAY		
1. Repair and maintenance	$ 26.60	None
2. Depreciation	$ 23.30	None
Subtotal	$ 49.90	$ 0.00
C. CLAMP EXPENSE PER DAY		
1. Added maintenance	Base	$ 1.70
2. Added depreciation	Base	$ 3.50
Subtotal	$ 0.00	$ 5.20
D. SPACE EXPENSE		
1. Pallet Storage Space**	$ 9.60	None
TOTAL VARIABLE COSTS	$111.70	$46.00

ANNUAL OPERATING EXPENSE DIFFERENCE
Clamp vs. Wood Pallet $16,500

*The figures used are for illustrative purposes only.
**Empty pallet Storage Space — Wood Pallets only. $.029/pallet handled x 333 pallets.
Assume: 20 pallets/stack occupying 30 ft.^{2} or 1.5 ft.^{2}/pallet
" Warehouse space value = $5.00/ft.^{2}/yr.
" A pallet turnover 12 times per year or every 4.2 weeks or every 21 days. THUS one out of every 21 pallets is in temp. storage.

$\therefore$ ($5.00 x 1.5 ft.^{2}) = $7.50 x $\frac{1}{21}$ or $.35 annual cost/pallet handled

$.35/yr. x $\frac{4.2 \text{ wks./turnover}}{52 \text{ wks./yr.}}$ = $.029/turnover

$.029 x 333 Unit Loads = $9.60/day.

FIG. 11-16. Cost comparison of pallet and clamp handling can be made easily by means of the form and formula used here. Courtesy of The Procter & Gamble Company.

The Unit Load, Pallets and Packaging | 171

Fig. 11-17. The size of package and container is increasing. Courtesy of Allis-Chalmers

12 | Storage Racks

Many owners of industrial lift trucks use some sort of storage rack system. Often the cost of the racks will be many times the price of the truck servicing them. Therefore, it is imperative that as much care be taken in planning and selecting the rack system as in selecting an industrial truck. The layout of aisles in rack designing is an especially critical factor in lift truck selection.

The easiest method in laying out a racking system is to do so before the building is constructed. However, most racking systems are installed in existing buildings and certain basic information will have to be gathered before making a layout.

If the rack is installed in an existing structure, the distance from the floor to the lowest obstruction hanging from the ceiling should be obtained. Usually, the determining factor in figuring out ceiling clearance is the sprinkler system. Most insurance companies require a minimum clearance from 12 to 36 inches from the top of the load to the sprinkler head. However, because the allowable clearance is often more than 12 inches, insurance carriers and local fire safety codes should be checked for the correct clearance. Figure 12-1 shows rack clearances.

Other service pipes, such as air, steam, water and electrical lines, light and heating fixtures, and/or structural beams or fire drops should also be measured for proper clearance. If a building structural beam is the determining obstruction, check the deflection for clearance as well. If there is a floor above, or in the case of a snow load on the roof, building deflection could be enough to cause an interference.

Another building factor to check is the floor loading. In mill-type multistory buildings the floor loading may be as little as 100 pounds per

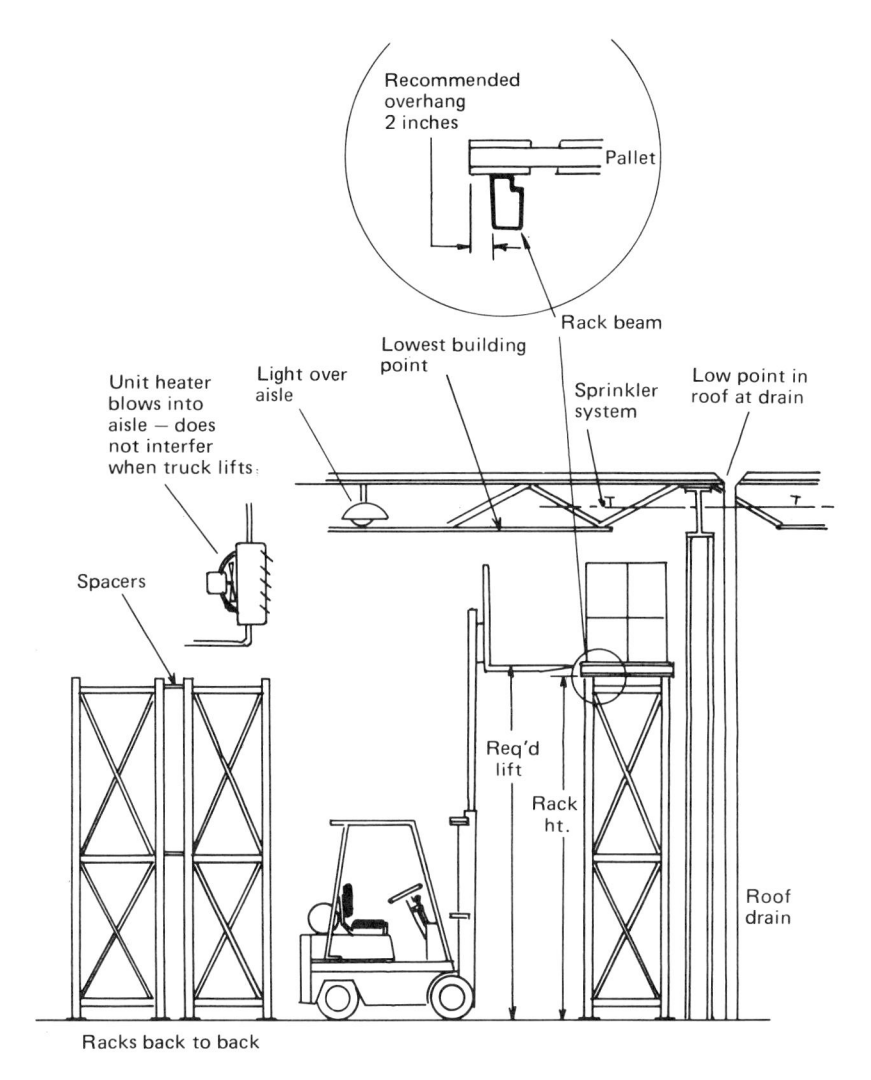

Fig. 12-1. Proper clearances are most important in rack installation. Heaters, lights, and other utilities should be placed above or away from the space required for a raised truck lifting mechanism.

square foot. With low floor loadings, a substructure may have to be built under the rack to distribute the loading on the columns in the rack structure.

Other building factors of importance are column locations and heater locations. In the layout of the rack structure, a column could end up in the middle of a proposed aisle. Heaters may affect the contents of certain products and may have to be relocated. Naturally, lights should be over aisles.

Equally important to the building data are the characteristics of the load. In most cases the load will be a unit load placed on a pallet or skid. The skid dimensions are important because the skid comes in direct contact with the rack. Pallets or skids may also require that special channels or other provisions be built into the rack.

Even though the load is on a pallet, its dimensions should also be noted. Any part of the load that extends beyond the pallet can be damaged. Load weights should also be noted. Even though racks have a built-in safety factor, it is possible that all the heavy loads could be placed in the same rack bay, causing potential damage to the structure.

Other load data required are the quantity to be stored, the turnover frequency, and the variety of loads. If large quantities of a unit load are being stored, then a drive-in rack might be used. If the mix or variety is high and order picking necessary, then a standard pallet rack might be a wiser choice.

An important factor in the load characteristics is the load's shelf life. If the cartons absorb moisture, over a period of time they may lose their strength and become a safety hazard. Because of possible product deterioration, the system may have to be first-in-first-out.

There are two main factors in laying out a racking system which are greatly influenced by the truck: the width of aisles and the height of the rack structure. Naturally a side loader will use a narrower aisle than a standard counterbalanced truck. However, the counterbalanced truck is more maneuverable. No matter which truck is used, additional maneuvering room is required for driver error when laying out the aisles. For example, the load may be "cocked" on the truck's forks or the load may not be placed completely in the rack. Several inches should be added to the aisle width to compensate for a cocked load. Then too, there is the question of speed. The more tightly the aisle is held to the minimum clearance, the slower the operation will be.

Because of the lift truck's design, the higher a load is lifted, the lower is the margin of safety between a stable and an unstable condition. The visibility of the driver is also lessened. Since a pallet is often used in the racks, the actual load lift height is from the floor to the bottom of the pallet and not to the bottom of the forks. If the low point of the building's ceiling is not the sprinkler system, then a minimum clearance between the top of the load and this obstruction should be at least 6 inches.

Of course, the determining factor in any rack system is comparing time against space savings. The higher the rack structure, the lower the building cost in dollars per square foot. However, the higher the loads, the more time required by the truck operator. The latter situation may require adding another truck and its yearly operating expense to the plant's truck fleet.

Standard Adjustable Pallet Racks

Probably the most common rack design is the standard steel adjustable storage rack (fig. 12-2). A standard rack consists of a pair of upright end frames and beams. The beams are spaced to support the front and rear end of pallets or to support a shelf or some other load sup-

FIG. 12-2. The system using the standard steel adjustable storage rack can be built to store almost any conceivable weight up to almost any height. Courtesy of Allis-Chalmers

porting device. Because the beams usually have connectors built into their ends, the beams are easily attached to the upright frames. The end frames are usually punched to allow the beams to be adjustable on either 2-, 3-, or 4-inch increments.

The previously mentioned rack structure is referred to as a rack bay. To add another bay, another upright spaced by beams is added to the structure. Standard racks are usually placed back to back in warehousing operations.

Since the pallet will be resting on the beams, a slight overhang is permissible. Acceptable overhang is 4 inches, with 2 inches being the normal recommendation. Overhang enables quick positioning of the pallet. However, any overhang allowances should be added to the aisle width when laying out the racks.

Because the upright frames take up space, the beams may be designed to hold more than one load. This type of design, however, does require that the beam be much thicker, which in turn may raise the height of the rack.

Minimum clearances around the load in the rack structure are 3 inches on each side. If there are two loads on one shelf, there should be an additional 4 inches between loads. Therefore, when determining the height of a rack, the height of each load (distance from the top of the load to the bottom of the pallet) will be added to the thickness of the beams plus the clearances required for each beam opening. To determine the overall height of the rack system, the top load should be added along with its recommended clearance of 6 inches (or, in the case of a sprinkler line, 12 inches or more).

Because of the overhang of the load, when racks are placed back to back, a spacer will be used to separate the two structures. Actually the spacer will tie the two units together, forming a stronger structure.

In a standard rack structure, unit loads are usually placed at the top and broken or open loads for order picking are placed at the bottom. If unit loads are stackable, two loads may be placed vertically on a beam. This procedure conserves space as a beam thickness and clearance is eliminated from the total rack height.

When specifying beams and uprights, a layout should be made showing the exact load requirements. This layout not only shows the exact dimensions of the rack, but also the weight limitations.

The beam of the rack may be constructed of angle iron, I beams, channels, pipes, specially rolled steel tubes and shapes, or wood. The present tendency in beam construction is to roll a special beam, which allows more flexibility to the shelf structure. A ledge is rolled into the beam, allowing placement of shelves, crossbars, and skid channels.

Standard racks are usually used when incoming stock is in unit load form and a high degree of selectivity is required, when a variety of unit loads are being received, or when outgoing loads are in broken lots. Fast moving loads or, as mentioned previously, broken loads should be placed at lower levels on the rack structure.

With the use of shelves or special accessories, the standard rack can store a variety of goods (fig. 12-3). Drum supports can be added to the beam, allowing drums to be stacked in a horizontal position. These supports are especially desirable in outside storage areas, because water then cannot collect in the top of the drumhead, causing rusting, destroying labels, or creating other serious problems. By using attachments on the forklift of the truck, unit loads may be placed on the shelving rather than on skids. This procedure conserves the overall height of the rack because the pallet or skid heights can be subtracted.

Even though most standard storage racks are adjustable, most of them are seldom moved. However, they can be put together easily and often unskilled labor can be used when installing such a system.

Flow Racks

By adding skate-wheel or roller conveyor sections to the beams of a standard rack, the structure can be converted into a first-in-first-out system. The conveyor is pitched so that goods flow from the back of the rack to the front by gravity. Care must be taken to prevent the loads from picking up momentum, because of the pitch, and damaging the rack structure. If the pitch is too slight, the loads may hang up. Braking devices, which allow easy movement to the loads, yet prevent too fast a build-up in speed and damage to the rack, are available.

In a flow rack, pallets or plywood boards may have to be placed under the loads. As in the case of cardboard boxes, the rollers will indent the bottom of the box, preventing its movement. Proper bracing is the most important consideration when a flow rack is used. Because mov-

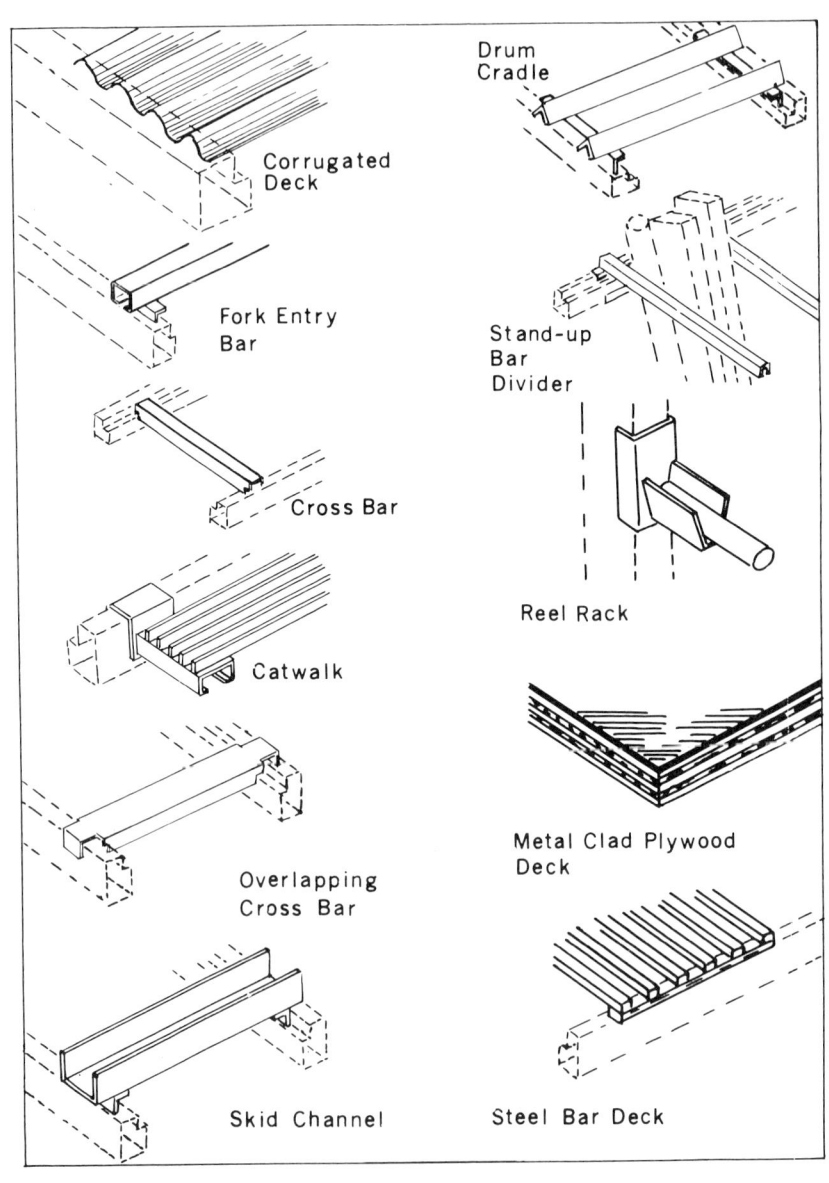

Fig. 12-3. Special rack accessories allow storage of almost any type material.

ing and impacting loads are constantly placed on the structure, the rack is subjected to many stresses not encountered in a standard rack structure. If the system is unusually large or high, it would be advantageous to call in a consulting service for the structure design.

Drive-in and Drive-through Rack Structures

In large storage systems, full utilization of the cube is most important. Quite often in such situations, loads are repetitious and uniform in size. As a result, loads can be stored in depth as well as in height. In other words, one aisle can store many loads deep.

If the load is a first-in-first-out type, the rack structure should allow loading from one end and unloading from the other. The rack used in this case would be a drive-through structure (fig. 12-4a).

If the load is a first-in-last-out type, unloading and loading is done from the same side of the rack or aisle face. The rack for this situation would be a drive-in type (fig. 12-4b).

In both instances, a fork truck can drive into the rack structure and deposit the load on a continuous ledge fastened to the columns. This ledge runs the entire depth of the rack and adds strength to the structure.

The difference between the two structures is in the vertical bracing. A drive-in rack is braced at the end of the rack (which prevents the lift truck from driving all the way through). A drive-in, drive-through structure develops its rigidity through overhead bracing either at the top of the rack structure or through tying into the building structure overhead.

The advantage of drive-in or -through racks is that one aisle is serving many loads deep (fig. 12-5). Efficiency can be stated as rack area divided by total area or $E = RA/TA$.

As an example, if we had a warehouse that was 56 ft wide and 144 ft long under a simple back-to-back rack system and were using 4-ft pallets and 8-ft wide aisles, the layout could look as in figure 12-6. The efficiency would be 3328/8064 or 42%. However, if the rack structure used was a drive-in type, the efficiency (see fig. 12-7) would be 6812/8064 or 85%. Of course, this is an aproximate figure because no figure was added for columns, widths, clearances, and other factors in the rack structure. Even so, one can easily see how the layout can greatly affect

180 | Lift Trucks

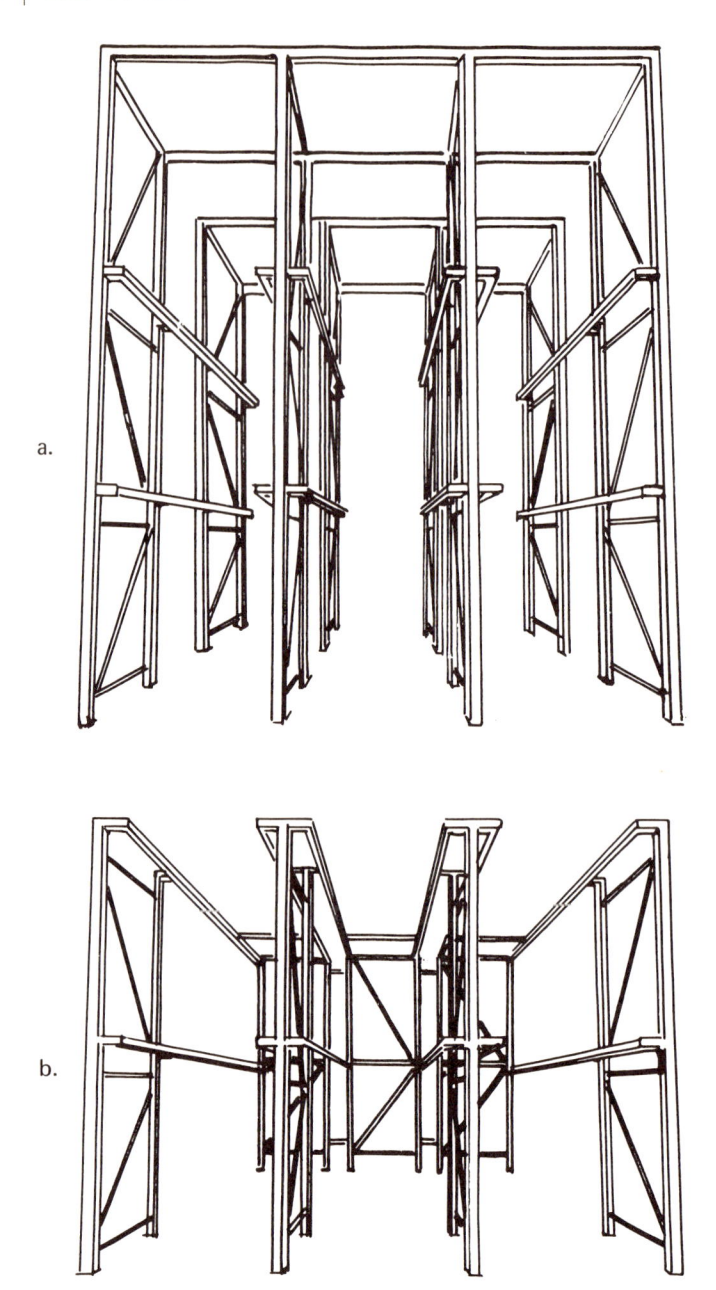

FIG. 12-4. A drive-through rack (*a*) allows first-in–first-out storage while a drive-in rack (*b*) is first-in–last-out.

Storage Racks | 181

FIG. 12-5. Either drive-in or drive-through rack systems eliminate many of the aisles required in a standard rack installation.

182 | Lift Trucks

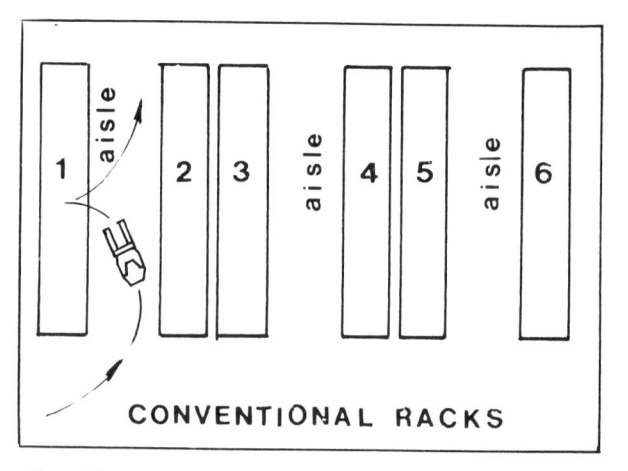

FIG. 12-6. Standard racking uses considerable aisle space, although it does take advantage of the building's height.

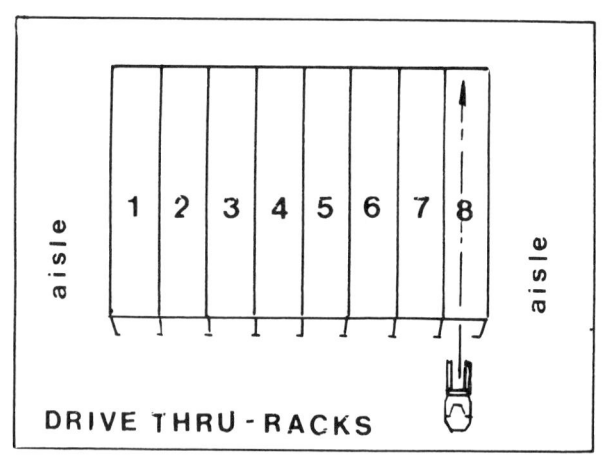

FIG. 12-7. The drive-in/drive-through principle is summarized in this illustration. As shown, many aisles have been eliminated.

the space utilization of the building. The cube factor is brought into the picture by considering the amount of loads stored vertically. If the floor to ceiling height is 24 ft, then the cubic efficiency would be

(total cubic load stored) / (cubic content of the building) .

Naturally there are limitations when figuring rack space. If the product has a high turnover, then perhaps a drive-in or drive-through structure would not be feasible. However, by changing the type of truck used and choosing a narrow aisle truck instead of a standard truck, the efficiency can be improved. Excluding normal clearances, suppose a truck could operate in a 4-ft aisle: another complete row of standard racking could be added.

If perhaps the forks could be 8 ft long and products stored two deep, then two complete aisles could be added to the system.

Of course, there are other considerations when comparing layouts. If the aisles are made too tight, a longer time will be used by the truck when maneuvering in the rack system. In industry, time means money and money has to be accounted for. The end result may require two trucks versus one under a less efficient layout.

Cantilever Racks

The cantilever (or sometimes called a Xmas tree rack) consists of a column with arms cantilevered off the column (fig. 12-8). Loads may be placed directly on the arms or on shelves supported by the arms. Frequently, because the loads placed in a cantilever rack would be quite long, this would be a good area in which to use a side-loading lift truck.

Rack columns may be self-supporting or braced to the ceiling at the top or to another rack. One rack manufacturer has constructed the arms so that they pivot upward off the column. The construction prevents the arm from being damaged if a fork tine hits it during a lift.

Honeycombing

For efficient storage, one should minimize losses in aisles and losses from vacant space in storage blocks. The amount of aisle space required in a storage area can be reduced by increasing the depth of storage from the aisle. However, by increasing the depth, the turnover rate of the goods in storage may create blocks of empty space in each storage bay. These losses in storage space are called *honeycombing*.

Honeycombing may be vertical or horizontal. As mentioned, horizontal honeycombing results from having partial bays. Vertical honey-

184 | Lift Trucks

Fig. 12-8. Cantilever racks (also called Xmas tree racks) allow the storing of bars, tubes, or other long items. Courtesy of Ingersoll-Rand Co.

combing results from having partial stacks. It should be noted that both are caused by the same conditions. Either the original order did not fill the row or stack, or empty space is created as a result of goods being removed faster than they are replenished.

In horizontal honeycombing problems, the depth of the bay can be reduced to compensate for too much vacancy. Vertical honeycombing can be reduced by using rack storage, especially when small lots are being handled.

Rack Buildings

A recent development in large warehousing operations is to have the rack frame support the walls and ceiling of the building. Such systems are actually structures and should be designed according to local building codes.

Storage Racks | 185

The building is a single-purpose structure and can be permanent or temporary. However, such buildings are quickly assembled and would offer some cost savings. Areas for accumulating goods might be rather limited because long spans would not be available. However, a separate structure could be built for incoming and outgoing load storage.

Other Rack Systems

There are various rack components for changing a pallet into a self-stacking unit. This type of hardware is covered in the chapter on pallets.

One other rack structure worth mentioning is the mezzanine (fig. 12-9). In high-ceiling areas it may be desirable to have an area for the storage of lightweight goods. By placing a mezzanine between rack frames, an area for light storage is made available. The lift truck can be used as an elevator to bring goods into and out of this area.

FIG. 12-9. A mezzanine can also take advantage of the space over aisles. Courtesy of Palmer-Shile

13 | Shipping and Receiving Docks

Frequently, the most neglected areas in industrial plant design are the receiving and shipping docks. It is at this point in the flow of materials where goods will really be expedited or all handling gains in the production and storage areas will have been in vain. Lift trucks are used in this area and safety is a must. As a result, the correct dock board selection can be an important factor in good lift truck operation.

The most important rule in dock design is to treat the truck, rail car, or whatever carrier is used, as an extension of the plant. Many dollars can be saved in the handling process by analyzing goods as they enter and leave the plant.

For example, a plant in the Midwest installed a monorail system in the truck trailers serving the plant. When the trailer was backed into the shipping dock, its monorail attached to a monorail system in the plant. Goods were conveyed right into the truck without any packing or double handling.

However, this efficient system is the exception and not the rule. Most loads are not that easily handled. They are usually in some form of a unit load and handled by lift truck or other means. Nevertheless, there are certain dock designs that will facilitate the movement of goods and they should be noted on the plant layout.

Location of the docks is of prime importance. If the trucking area is outside, the prevailing wind direction may greatly affect the climate within the building. Every time the doors are opened or closed, temperatures in areas adjacent to the dock will rise and fall.

Another factor influencing location would be local zoning codes. There is the story of the plant that was located in a suburban area, where trucks made their approach through a beautiful housing development. Needless to say, the irate homeowners forced the company to change the layout of the building and a new access road had to be provided.

Further considerations in location are the volume of traffic, types of trucks delivering and unloading goods, and whether outdoor loading and storage is required. If outdoor unloading and storage is necessary, a ramp should be convenient to the dock area allowing lift trucks to go from the plant into the yard.

Coupled with location in the overall problem of dock design is the style of the building which houses the dock area. Traffic flow may dictate that there be no central dock area. For instance, an eastern steel warehouse has three main aisles running the length of the building; one along each outside wall and one through the middle (fig. 13-1). All incoming goods are delivered along the outside walls and all shipping is through the middle aisle. Goods flow from the outside walls to the middle aisle. Most loads arrive on flatbed trailers and can be unloaded by lift truck or crane without the use of docks.

Other building styles include the partially enclosed and totally enclosed dock areas. These styles vary from buildings with only canopies to cover the trailers, to dock systems in which the truck drives into and does all maneuvering inside an enclosed area (fig. 13-2 and 13-3).

The dimensions of the dock are most important and two major considerations are: First, the space per truck spot and, second, the number of truck spots, also referred to as berths.

Rather than providing a minimum maneuvering area for the trucker, the truck berth should be as roomy as possible. This roominess will prevent the considerable minor damages that can occur when spotting the truck in a tight space. A berth can be 10 to 14 ft wide although 12 ft is the average. The length of a berth will be determined by the apron space and the presence of any other trucks or obstacles (fig. 13-4).

The apron space is the minimum distance required by the truck in backing up. This distance can be determined graphically. However, with the variety of trucks entering the dock area, the data in table 13-1 should be useful.

The number of truck spots or berths is determined by the workload of the dock area. More specifically, this number can be determined by

188 | Lift Trucks

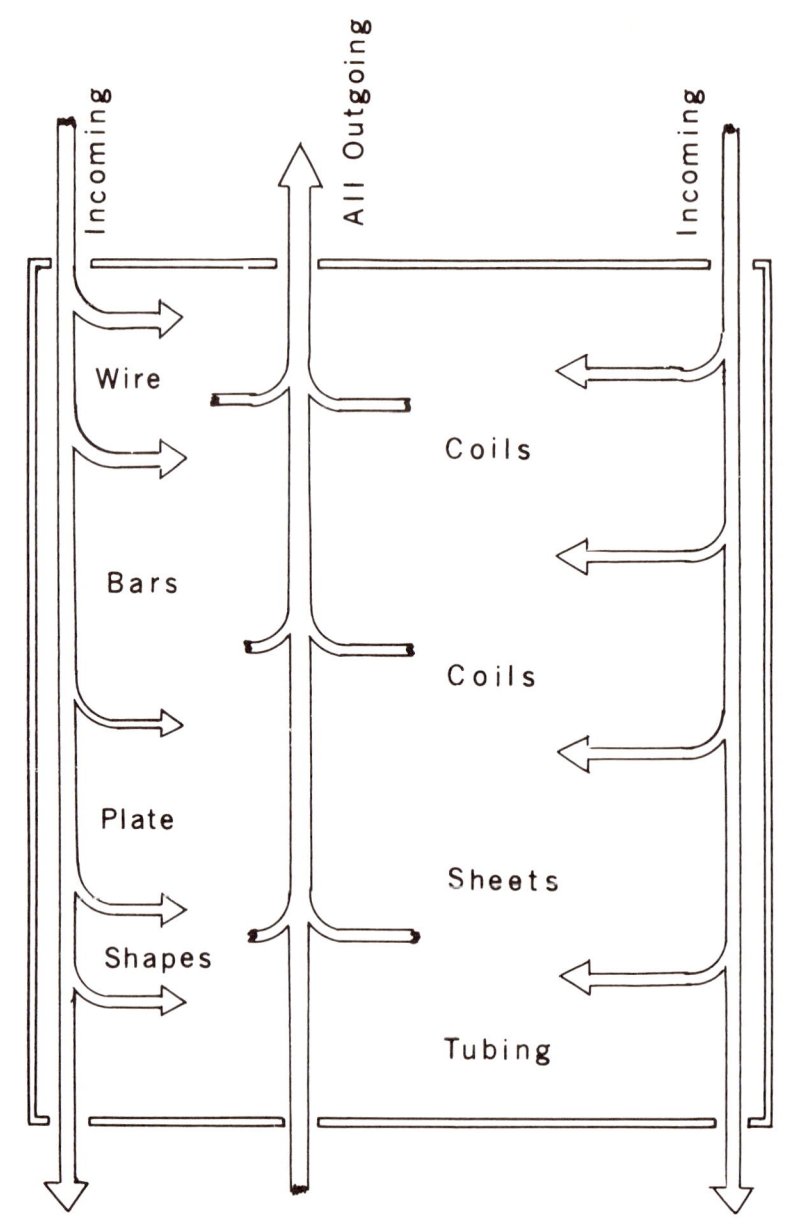

FIG. 13-1. In this plan of warehouse without a dock, three main aisles run its entire length. Materials are unloaded at outside walls and flow to the center aisle for shipping. Aisles are wide enough for two over-the-road trucks to pass one another.

Shipping and Receiving Docks | 189

Fig. 13-2. Canopy-protected dock prevents rain from making dock area slippery.

plotting the amount of time necessary to handle each truck against the arrival patterns of the carriers. Usually, the peak period is used when figuring truck spots. Therefore, if the peak period is 10 trucks per hour and it takes $\frac{1}{2}$ hour to handle each truck, the number of berths required would be $\frac{1}{2} \times 10$ or 5 berths.

If the dock activity has great extremes, it may not be desirable to use a peak period for determining the number of berths. Simulation methods may be used instead to calculate a proper estimate of the number of berths required. However, in large dock installations an engineering firm will probably be used in the design and this probably would not be a concern of the average reader.

Outdoor considerations in dock design are drainage, paving, lighting, communications, and, where applicable, snow removal. Yard drainage should be away from the dock area. Water or snow may cause poor traction. Unpaved lots create many problems. Potholes form in bad

190 | Lift Trucks

FIG. 13-3. Completely enclosed dock area eliminates any weather problem; it also allows maximum security. Courtesy of Poweramp

weather, causing difficulty when maneuvering the trucks. During dry spells, these lots can become quite dusty and the dust created will usually end up in the plant.

A further advantage to paving occurs in situations where outside loading or storage takes place. Most lift trucks in the plant will have difficulty traveling over gravel lots unless they have pneumatic tires.

Paving should be designed for local conditions by a competent engineer. Usual load limitations are 18,000 lbs on a single axle and 32,000 lbs on a tandem axle. If the truck disengages from the trailer, extra pavement may be necessary where the landing gear rests. If this is the case, a strip of concrete under the landing gear will prevent the trailer from pressing into the ground or asphalt.

Since every product produced by the plant will be passing through the shipping dock, it is wise to have the outside of this area well lighted and protected by a fence and even a guard system. Other safety and

Shipping and Receiving Docks | 191

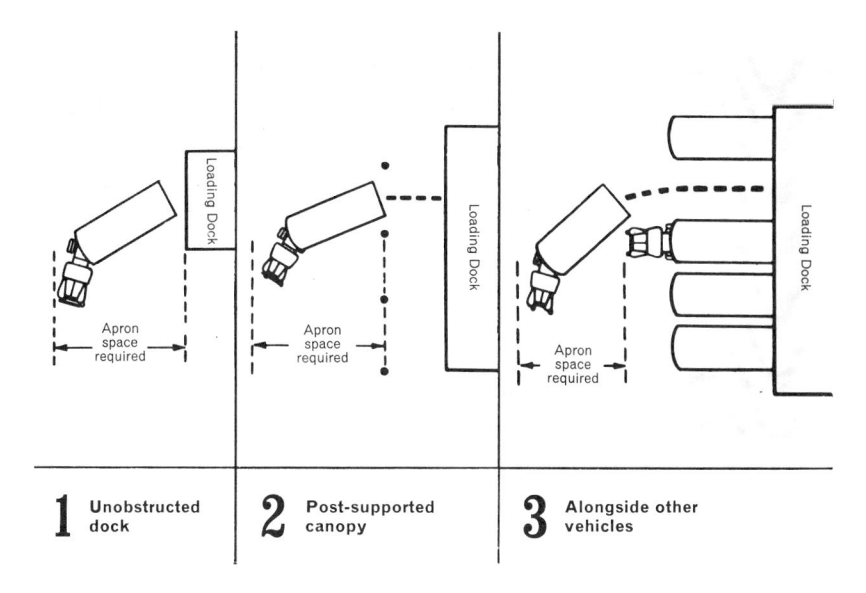

FIG. 13-4. Maneuvering space required by over-the-road trucks at a loading dock depends on several factors. Courtesy of Kelley Company, Inc.

Table 13–1. Apron space requirements (in ft)

Length of tractor and trailer	Truck spot width		
	10 ft	12 ft	14 ft
40 ft	46	43	39
45 ft	52	49	46
50 ft	60	57	54
55 ft	65	62	58
60 ft	72	69	63

security measures include yellow guide lines painted on the pavement and wheel chocks to prevent trailers from moving. If there is a great deal of traffic, a special communications system may be required to speed placement of trucks. This system may include a telephone location for incoming trucks as well as a closed circuit television system.

If yard communications become necessary, they could be handled by connecting an outdoor speaker to the plant PA system. Any fencing in the yard should be well protected. Curbs should be at least 5 ft from

192 | Lift Trucks

the fence and if timbers are used, they should be well secured by pipes driven into the ground.

The dock itself should be well protected against damage from trucks and from adverse weather. Bumpers should be used at the dock edge to lessen the impact from a backing truck (fig. 13-5). Laminated or molded rubber bumpers are recommended over timbers. They look better and have better impact absorbency, while timbers deteriorate and have to be replaced periodically.

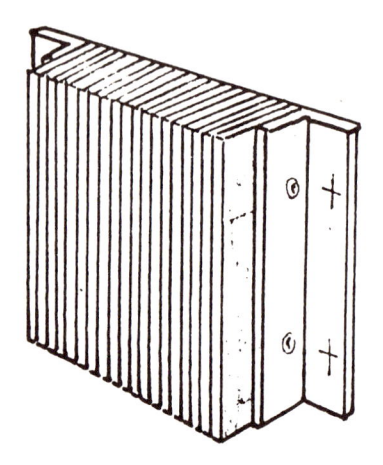

FIG. 13-5. Rubber dock bumper offers more protection than timber style bumper. This style absorbs shock from backing truck, whereas wood units transfer shock directly into the building.

If the dock is not inside, weather can be quite a problem. If a canopy is used, it should be at least 14 ft high. If there is a grade in front of the dock, the canopy should be higher yet. Also, the dock should extend away from the building if the driveway is depressed. Otherwise the top of the trailer may extend beyond the truck floor and strike the building wall.

If there is no canopy, some form of weather seal should be used at the doorway. Not only will wind or cold be a problem, but also, without a seal, water could cause slippery conditions when lift trucks are entering the trailer.

The height of the dock will vary considerably, depending on the trucks servicing the plant. Recommended height for pickup trucks is 48 in. and a 52-in. height for tractor-trailer combinations. Bed heights of trucks may range from 38 in. to 66 in. (most extreme cases). Some kind of leveling device will be required if lift trucks are used for loading and unloading the trucks.

Three basic styles of leveling devices are: permanent adjustable boards, portable dock boards, and a truck leveler. The permanent board is an adjustable platform attached to the building. It may be built into the dock itself, but there are also styles available that can attach to the front of an existing dock. The adjustment may be mechanical or manual (fig. 13-6).

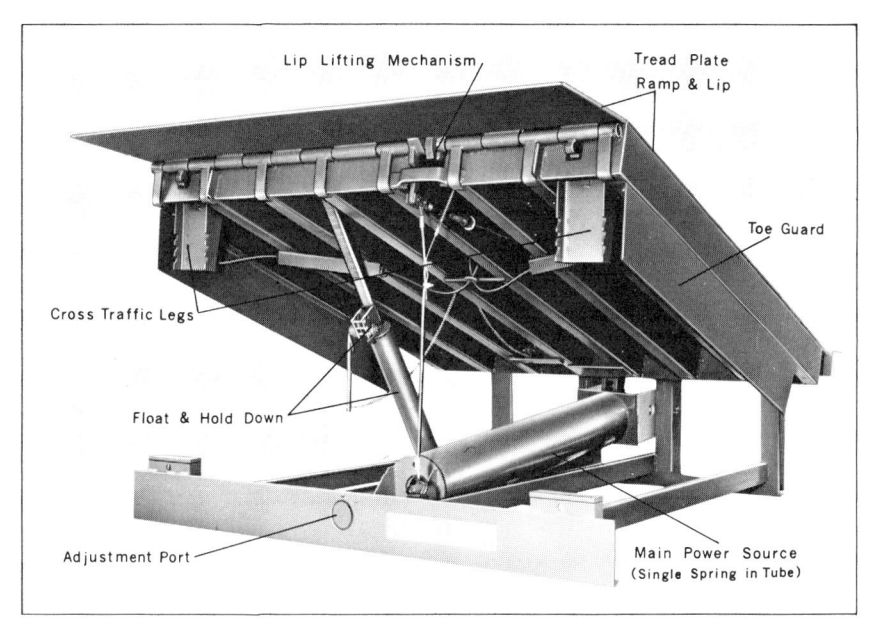

FIG. 13-6. Components of typical permanently installed dock board. The advantage of this type of board is that it cannot slip out from under a lift truck when the lift truck is entering a trailer. Courtesy of Kelley Company, Inc.

The portable dockboards are usually made of lightweight metals and are less expensive than the permanent boards. However, they must be brought into place each time a truck is loaded.

Truck levelers are built into the bottom of the truck pit and raise the rear end of the trailer so that its floor is level with that of the shipping dock. They are quite expensive and seldom used.

If there is a high volume of traffic, the permanent board is the most desirable. It is safer than the portable board and cuts loading time. If a portable board is used, it should be purchased from a reputable manufacturer and not "home" built. The board should have curbs, position locks, safety stops, and proper load capacity. The use of smooth steel plates is a dangerous policy and is not recommended.

14 | Yard Handling

Ingenuity is the key to becoming a successful industrial engineer. It is also the key to solving many storage problems. With the growing need of space, many plants are turning to their yards for storage. Some of the problems in yard storage, as well as tips on outside truck operation, are presented in this chapter.

The old saying, "There is more than one way to skin a cat," certainly applies to yard handling. Counterbalanced trucks, sideloaders, yard cranes, and straddletrucks may all be considered for the same job. How does the materials handling engineer decide on which unit to select for his particular operation?

First, the engineer must realize that, in order to have an efficient operation, outside handling may require as complete a handling analysis as an inside handling job. For example:

At a plant where I once had worked, lumber was received periodically for a crating operation. Originally, lumber cut in random lengths was received in boxcar loads; unloading was done manually at the shipping dock. A skatewheel conveyor carried the lumber outside, where again it was handled manually — this time to stack it. From a cost-per-board-foot standpoint, this was a very cheap way to buy lumber.

However, when adding the price of the material to the labor to handle it, plus the cost of the scrappage created by random lengths, the lumber was not cheap. A cost analysis showed that it would be much more economical if the lumber was purchased in precut lengths, shipped by truck, and then unloaded by lift truck.

At the time it was thought that by grading and covering the lumber storage area with gravel, in-plant solid-tired lift trucks could then be used to hoist the unitized loads off the over-the-road truck's trailer. Needless to say, downtime due to trucks being stuck in the gravel became quite excessive. Not only the truck operator, but usually several maintenance workers as well, were tied up. In addition, production in the press room could also be tied up, waiting for the use of their lift truck.

What had seemed like a simple materials cost analysis had mushroomed into a traumatic problem. The shipping department foreman would become angry as his workers would leave their jobs to watch the operation. The maintenance foreman would become equally angry as his men were pulled off scheduled jobs and forced to work on an operation that could be quite time-consuming. In addition, the maintenance workers would usually then have to work overtime to finish their own scheduled work. The pressroom foreman would then add to the confusion and become "raving" mad, because it was his department's truck that was being used as it was the only truck in the building that could lift this particular size load.

An investigation was started to see just what could be done to resolve the situation. Would it be feasible to blacktop the area so the solid-tired truck could be used, or should we buy a pneumatic-tired truck and forego the maximum utilization of the vehicle?

As it turned out, neither suggestion was accepted. During the investigation, someone mentioned that there was a machinery mover, with a pneumatic-tired truck exactly the size we needed, available nearby. All we had to do was set up a simple truck rental program. The program chosen included a machine operator and, even though the owner was paid a premium price, it was successful and everyone in the plant was happy. No longer was the pressroom lift truck tied up, no longer were maintenance workers called on for unscheduled work, and, best of all, the shipping department foreman got his lumber truck unloaded on time. In return the entire operation was more economical than before and it was accomplished without making any capital investment.

Product Analysis

An analysis of the materials handling in the yard should start with a product analysis. Some of the points to look at are weight of the product, length, fragility, and whether or not the material can be tiered or unitized.

All materials that must be stored outside are subject to the effects of weather. It might be cheaper to store the material inside a building rather than outside; if it is stored outside it may have to be derusted, as in the case of steel, before it can be used.

When analyzing a particular part of a truck, consideration should be given as to how the part fits into the overall material flow. For example, suppose the part is important to an assembly process and, without it, production would be stopped. If, during a snow storm, the part was not obtainable, then it would be advisable to either store it inside or have a sufficient number of parts in an inside storage bank for at least one day's operation.

Identification

Not only must the part be able to withstand the effects of the elements, its identification mark should also be capable of withstanding weather. Probably the most frequently encountered identification problem is found in the storing of drums. Rain can collect in the drum heads and destroy their labels if the drums are stored vertically. If the drums are stored horizontally in racks, water cannot accumulate and this problem is eliminated.

Identification can be carried out either by using weatherproof marking materials, or by placing the materials in mapped locations. This, however, requires a well laid-out yard and the driver must follow instructions to a "T" when storing items.

Terrain

The lay of the land will greatly affect any yard-handling analysis. If the area is paved, then solid-tired trucks may be used. However, if the yard is dirt or gravel, then pneumatic-tired equipment is necessary.

Weight of the product to be stored should be considered. At one particular steel plant, the weight of the steel forced the dunnage under the load into the ground, thus preventing the forks from getting under the load. Blacktopping the area was not recommended because, during hot weather, the blacktopping could become so soft that the steel would still drive the dunnage into the ground. Finally, concrete was poured in the areas where these large loads of steel were to be stored and it proved to be successful.

Important factors to consider in analyzing the terrain would include proper grading for water drainage and eliminating any steep ramp-like grades that could cause the truck to work too hard. Steep grades can also cause the load to tip.

Most yard vehicles will travel at high speeds (some having a 30-mph speed capability when not carrying a load), therefore, the area should be properly posted and have adequate aisles. Pedestrians should be kept out of the area and the vehicles should be equipped with horns and lights.

Railroad tracks are a problem in many yards. Unless the yard has a hard surface, extreme caution should be taken in using rubber-tired vehicles to move railroad cars. First, many vehicles are not designed for this function and, second, the torque of the truck's tires can rotate the tie from the underside of the rail. In addition, the loose ballast will act like ball bearings and cause a loss of drawbar pull. When figuring railroad drawbar needs, 20 pounds of drawbar pull are required to move one rolling ton of railroad cars. Because of the friction of the railcar wheel flanges, 2 pounds of drawbar are added for each degree of curve. And, for every one percent of grade, an additional 20 pounds should be added.

Car-moving operations should be cleared with the company controlling the rail movements. Only properly trained workers should move rail cars. One car out of control can be an expensive proposition, not only to the physical plant but, most importantly, to the worker, if he should be injured.

Equipment for rail car movement ranges from mechanical pinch bars to radio controlled locomotives. The type of equipment to use will naturally depend on how many cars must be moved, the frequency of moves, and for what distance.

One simple method is towing the rail cars with a cable puller. However, a cable puller is limited to a small area and usually confined to a straight track.

Roadable units include vehicles that ride on the tracks. One such unit has two sets of wheels. A set of steel flange wheels is used for track operation and a second set of rubber tires can be lowered for road use. Another type of trackable unit is a rubber-tired vehicle that has steel flanged wheels to keep the vehicle aligned on the tracks. For road use, the steel wheels are raised.

Other roadable vehicles are front-end loaders or pneumatic-tired lift trucks with a rail car coupler attached. As mentioned previously, a hard surface should cover the rail ties. Roadable vehicles should not be used unless they have the coupler. Pushing a car with a bucket or forks is extremely hard on the equipment. Even if the back of the vehicle is used for pushing, the rail cars are not under positive control and are quite dangerous.

Rail crossings should be constructed as smoothly as possible. Lift trucks should be driven across the tracks at an angle and as slowly as possible.

Yards require maintenance, including the filling of potholes and, in northern climates, snow removal. Aisleway in yards should be clearly marked and housekeeping enforced. In one brickyard application, a paved yard increased truck life 50% and reduced truck maintenance costs another 50%. In addition, each truck was able to move 50% more products each day. A well-maintained yard can repay every penny spent on it.

Lighting

Lighting not only allows the worker to see what he is doing, but offers security to the plant as well. Lighting should be designed so it does not blind a driver. Light poles should be located either away from the operation or be adequately protected by guard rails.

Headlights should be mounted both on the front and rear of the truck and, in some cases, a spotlight may also be needed.

Building

Often the building has to be considered at the same time the yard is laid. At one new marina installation the importance of layouts was brought out. Pleasure boats at the marina have to be placed into the water, then removed and stored.

Previously, this operation was very time-consuming as it usually re-

200 | Lift Trucks

quired a derrick and lifting cables. In addition, maximum utilization of floor or ground space was almost impossible as the boats could not be stacked.

The modern way to handle pleasure boats includes large lift trucks and storage racks. The trucks have forks that can be dropped below the ground level of the truck to lower boats into the water. The operation starts off with the truck lifting a pleasure boat out of the storage rack (fig. 14-1). The boat is taken outside and lowered into the water by extending the front of the mast off a pier or dock and lowering the boat into the water. The coordination of an inside operation with an outside handling operation pays off handsomely, as one operator can handle the entire job.

The following points are just some of the necessary requirements

Fig. 14-1. Forks that drop below ground level allow this lift truck to launch boats at a modern marina. Courtesy of Hyster Company.

that must be incorporated in a building that is part of a yard-handling operation:

1. Access to the building: Doorways must be large enough to accommodate the largest truck in the fleet. In fact, it may be desirable to motorize the door so the operator can open it without leaving the truck. This may be done through a ratchet relay pull cord or by radio control.
2. Visibility at the door opening: Usually a truck should be brought to a full stop before proceeding through a doorway. In addition, the operator may want to use his horn. In areas where the weather is a factor, an operator should be cautioned about the change in floor conditions. He may be going from a dry floor with good traction into light snow where there is very little traction.
3. Properly graded ramps which allow a truck to enter at a safe speed. A steep ramp which leads onto a level floor is inviting trouble, as the operator will have to brake his truck as he makes the transition from the outside to the inside.
4. Lighting should be arranged so that it does not blind the driver going from inside the building to outside, or vice versa.

Security

Just as lighting offers a measure of security, so does fencing around the yard storage area. Watchmen should pay close attention to these areas as all the items stored can usually be under the scrutiny of the general public. This is especially true in scrap operations where youngsters may want to sneak in to pick up scrap crating lumber and other miscellaneous items. Any items that the general public could be tempted to take should be protected by a solid fence that hides the operation from the public view.

Yard handling machinery should be equipped with keys to prevent its operation by untrained persons. Not only do children try to operate such equipment, but often an unauthorized worker may be tempted to give it a try.

Equipment Types

In addition to the counterbalanced truck, sideloaders, and straddle-trucks mentioned in chapter 3, two other vehicles often considered in a

202 | Lift Trucks

yard analysis are the crane truck (fig. 14-2) and the rubber-tired traveling crane (fig. 14-3).

The yard crane comes in many varieties and booms of different lengths are available for it. If a long boom is being used, it is extremely important that it be used carefully. Not only can the boom hit overhead electrical lines, but the boom will vary in lifting capacity according to its angle with the verticle surface.

A disadvantage of crane vehicles is that often a "rigger" or "hooker" will be required to attach a sling around the load to be lifted. However, the advantages of a yard crane can often outweigh its disadvantages. It does have the advantage of reaching over previously stacked loads to pick up material which would normally be impossible to pick up with a lift truck or sideloader.

Examples of how different vehicles can be used on long products are shown in figures 14-4a, 14-4b, and 14-4c. In figure 14-4a, a standard lift truck is shown lifting long steel beams. The tipping factor involved can

Fig. 14-2. This yard crane is a little different from most; however, it can reach over and select pipe from the back of the pile. Courtesy of Crutcher Resources Corp.

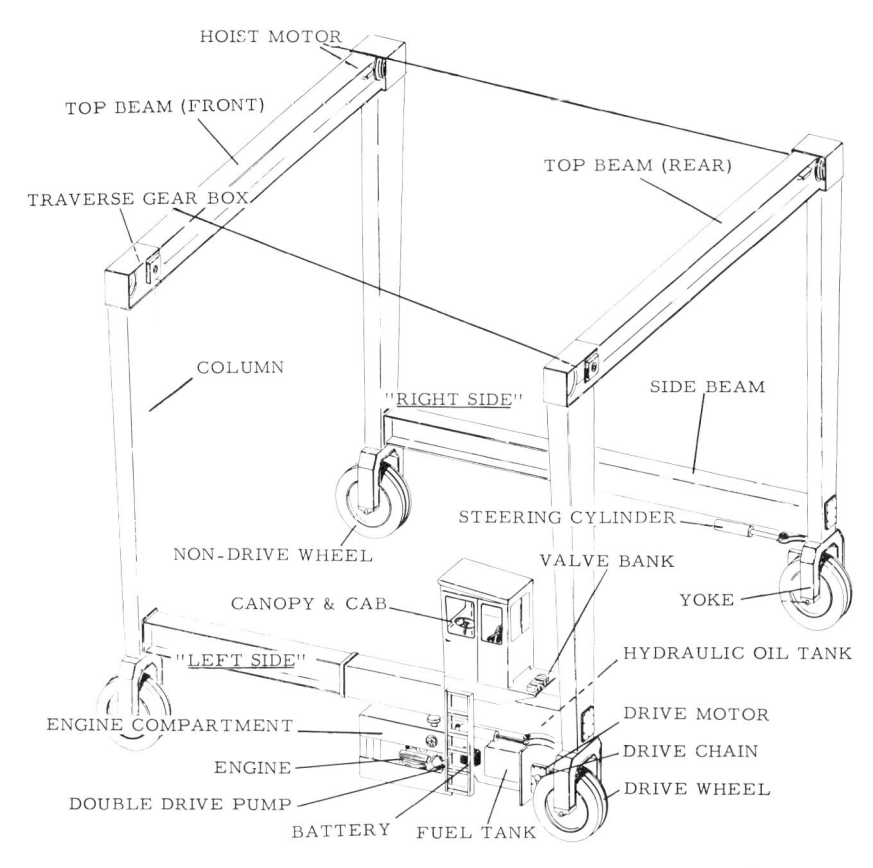

Fig. 14-3. Rubber-tired yard cranes provide great mobility in yard storage areas. Unit can be driven into the plant (provided its ceilings are high enough) and used as a replacement for an overhead crane. Courtesy of Drott

be easily seen and the truck should not be used regularly for this work — only occasionally.

Figure 14-4*b* shows a sideloader with the same type of load. Notice how the truck has a platform on which to place the load when transporting it. A compromise between the two other vehicles is shown in figure 14-4*c*. This truck is similar to a sideloader. It has a swivelling mast and can pick up a load much like the lift truck, rotate the mast, and carry the load.

Need will dictate equipment requirements and often a hybrid is necessary for a special application. The truck in figure 14-5 is a com-

204 | Lift Trucks

a.

b.

Fig. 14-4. Three different methods to handle the same load are shown in this series of photographs. A standard lift truck (*a*, courtesy of Hyster Company) is good for short hauls, but would have to run at slower speeds during a long haul to prevent spilling of the load. The sideloader (*b*, courtesy of Allis-Chalmers) has the speed required for outside handling. The swing front-end truck (*c*, courtesy of Matbro Sales Ltd, British Aircraft Corp, Inc), has the stacking qualities of the standard lift truck yet can swing the load on to its side similar to a sideloader.

bination of a front-end loader and the mast of a counterbalanced truck. The net result is a truck that has an exceptionally high lift. Notice the retractable jacks or outriggers that help stabilize the load. Jacks are used on various kinds of lifting equipment, including sideloaders and mobile yard cranes.

A variety of front-end attachments that make yard vehicles even more flexible is available. In the cement block manufacturing industry, a multitine fork attachment allows palletless handling of a unitized load of blocks (fig. 14-6).

Trucks need not be considered for production alone. For instance, the truck shown in figure 14-7 has a special attachment for lifting jet engines in a maintenance operation.

Containerized Handling

The container has become the prominent feature of a revitalized shipping industry. With the development of the containership came the development of vehicles to handle containers. With the prospect of more and more goods coming into the industrial plant via containers, yard handling will be developed even further. There is the possibility

c.

206 | Lift Trucks

FIG. 14-5. What is basically a front-end loader truck has a lift truck mast attached for greater lifting capabilities. Courtesy of Pettibone Mulliken Corp.

that warehouses could be eliminated and all products will be stored in over-the-road containers. After all, they are built like a small building and are completely weather tight. Some have their own refrigerating plants, eliminating the need for inside storage.

Yard Handling | 207

a.

b.

FIG. 14-6. As shown in (a, courtesy of Hyster Company), the lift truck can have a series of fork tines attached to the carriage for lifting a load of blocks. This technique eliminates the need for a pallet (b, courtesy of Clark Equipment Co).

208 | Lift Trucks

Fig. 14-7. This large lift truck has a front-end attachment that can carry jet engines. Courtesy of Hyster Company

Equipment needs will be determined by whether the containers are stacked and the length of time required for their transportation. Three different ways of transporting and tiering containers are shown in figures 14-8a, 14-8b, and 14-8c.

Construction Usage

Another form of yard handling would be involved in the construction of new buildings. The truck can transport materials on the construction site and, in addition, it can also serve as an elevator (fig. 14-9). Sideloaders can also be used in this way (fig. 14-10).

Many varieties of lift trucks are found in the construction industry, because a mast can be attached to almost any type of construction vehicle. One common vehicle is the tractor equipped with a bucket or scoop on the front end and a forklift mast attached to the rear.

a.

b.

Fig. 14-8. Three different methods for carrying large containers are straddle carrier (*a*, courtesy of Hyster Company), lift truck (*b*, courtesy of Towmotor), and sideloader (*c*, courtesy of Allis-Chalmers). The sideloader has a cab that can be used from either side of the truck, allowing the driver better visibility.

210 | Lift Trucks

Fig. 14-8c.

Fig. 14-9. Lift truck serves as an elevator at the construction site. Courtesy of Clark Equipment Co.

FIG. 14-10. Sideloader at a construction site. Unit can carry pallets as well as long lengths of lumber. Courtesy of Allis-Chalmers

However, lift truck manufacturers have made their vehicles most competitive by allowing them to be towed behind delivery trucks (fig. 14-11). This permits a worker to load his over-the-road truck with the yard lift truck, attach the lift truck as he would a trailer, and then tow it to the construction site where the same lift truck can then be used to unload the over-the-road truck.

Summary

Materials handling outside the plant can be just as important as handling materials inside the plant. By treating the yard as a standard facility layout, the materials handling engineer can make the outside operation just as efficient as the inside operation.

212 | Lift Trucks

Fig. 14-11. Towing attachment on lift truck allows it to tag along easily behind delivery trucks. Courtesy of Hyster Company.

15 | Applications: Problem Solving with Lift Trucks

"The proof of the pudding is in the eating" and the proof of the versatility of the lift truck is the many areas in which it is used. The lift truck revolutionized the manufacturing plant and gave management and the worker a tool that not only makes a worker efficient, but also makes entire operations economical. By observing some of the more unusual applications of lift trucks the reader may be able to solve some of his own problems.

Lift trucks will be found in almost every manufacturing and warehouse operation in the country. Often used on shipping and receiving docks, they also are used extensively in warehousing operations. Even in the automatic warehouses, the lift truck is needed to load and unload conveyors and storage/retrieval devices.

Versatility Still Key Feature

Similar to the marina operation explained in chapter 13 is the metal-cleaning operation shown in figure 15-1. Downward-extending forks were added to the standard counterbalanced truck, allowing the driver to lower materials as well as raise them.

The lift truck offers versatility even without attachments or special alterations. As shown in fig. 15-2, a lift truck can handle cylindrical objects as easily as it can handle box-shaped objects.

In addition to being a useful handling device by itself, the lift truck works well in conjunction with other handling devices. One example of coordinating the movements of one handling device with another

214 | Lift Trucks

a.

FIG. 15-1. A lift truck's versatility is shown by this lift truck in an acid cleaning operation (a). In (b) the tank has been drained to show how the forks can extend downward.

was proven in the following warehousing operation at a television manufacturing plant.

The operation consisted of an elevator, a conveyor bridge, a lift truck, and an overhead towline conveyor system. The elevator lifted boxed

Applications: Problem Solving with Lift Trucks | 215

b.

TV sets off the plant floor up to a conveyor bridge that crossed the railroad tracks. The boxes were automatically shunted off the elevator onto the conveyor, which carried them into the warehouse located on the other side of the tracks. The lift truck hoisted the sets off the conveyor, then the operator drove up behind the cart being towed by the towline, synchronized his truck's speed with that of the towcart, and skillfully placed the boxes on the cart.

216 | Lift Trucks

Fig. 15-2. By tilting the mast backward, a lift truck can carry cylindrical objects. Unloading is simple — just tilt the mast forward. Courtesy of Exide

The operation was a good example of making the most out of an extremely poor building layout. Because the railroad was being used constantly, it was impossible to allow trucks to cross the tracks. By using a conveyor, the bridge construction over the tracks was held to a minimum weight. With the span involved, it would have been prohibitive, costwise, to construct a bridge with sufficient capacity to carry the weight of a lift truck, as well as the load.

The distance to where the sets were placed in storage in the warehouse was too great for an economical lift-truck run. Likewise, the cost of an automatic operation to transfer the sets onto the moving tow carts was prohibitive.

Operating Characteristics

The lift truck is an extremely durable piece of equipment. If one were to compare the operating life of a lift truck with that of the average automobile, he would find that the lift truck outlives an auto several times over. In addition, the environment in which many trucks exist is almost unbelievable.

Foundry operations can be extremely harsh on the lift truck, yet it is surprising how many will last far beyond their projected life considering the environment in which they operate. Pressrooms in many sheet-metal plants can be particularly rough on lift trucks, but, by applying the following hints, truck maintenance will be kept to a minimum.

First, housekeeping should be enforced. Metal chips should be kept out of the aisles and any oils or greases should be immediately soaked up by an absorbent. Metal chips and oils are very hard on tires and oil, especially, is an operating hazard (fig. 15.3).

Aisleways should be well marked and they should be kept clear of obstacles. Both lift truck and production equipment damage can be held to a minimum by providing adequate maneuvering space.

Last, but not least, the operator should become thoroughly familiar with lift-truck operation through a comprehensive training program. Tight spaces require careful maneuvering and, without adequate preparation, an operator could easily cause damage to the truck, product, and, most importantly, to himself or a bystander (fig. 15-4).

These items may seem repetitious, but they are to be emphasized, as the lift truck can be easily abused. Far too often the lift truck is taken for granted. Because of the truck's durability and versatility it is easy to forget that the lift truck is quite a complex piece of equipment.

Compare the treatment given to a lift truck with treatment given to other equipment in the plant. Most production equipment will have

218 | Lift Trucks

FIG. 15-3. Cleanliness pays off in the long run. Loose steel packaging straps can cut and damage the tires and any oil spills should be cleaned immediately.

strict limitations regarding capacity, speed, and maintenance procedures and these requirements are usually followed. Yet, operators will use lift trucks to lift loads that raise the truck's steering wheels off the ground, or they will race the engine without any load on the motor. They will also use the truck to bump open doors, push machinery, and, in some extreme cases, drag race with other workers. Employees of almost every department in the plant can abuse the use of lift trucks. Foremen will beg off having the truck maintained because they want to use it for an extra hour's overtime, or maintenance men will stop repairing a lift truck and place it back in service so they can run out and maintain some production machinery.

Applications: Problem Solving with Lift Trucks | 219

FIG. 15-4. Tight space requires a well-trained driver. Exceptionally heavy load points out the need for carriage and overhead guards. Courtesy of Baker Div. of Otis

Dollar for dollar and pound for pound, the lift truck probably takes more abuse and gives more in return than any item in the shop. The lift truck is durable and with adequate care should provide many hours of dependable service.

Flexibility

Many items are purchased for one purpose only. However, because of its versatility, the lift truck is frequently pressed into service, performing operations it was never intended to do. The lift truck makes a good elevating work platform. Or, it can serve as an elevator, raising material from one floor to another (fig. 15-5). And, in many shops, a lift truck becomes an indispensable piece of equipment for the maintenance department.

Work platforms are often built to slip over the lift truck forks, which permit electricians and pipefitters to service overhead utilities. In such cases, the truck is far more reliable than a ladder as the maintenance worker not only has a platform to work off of, but tools and parts can be lifted up with him.

Flexibility of a lift truck is best shown when the route or location of a materials handling operation is determined. Almost every kind of handling device, except the lift truck, will require an extensive renovation program in order to perform more than one function.

In addition to its tiering ability, the truck is flexible enough to be used in a low-ceiling basement area (fig. 15-6).

Cold Storage Warehouses

Prior to World War II, most cold storage warehouses were low-ceiling, multistory buildings. Handling was limited to a central elevator and hand stacking. Commodities were handled on a seasonal basis, usually received at one time. Few items were frozen and products were withdrawn in small quantities during the year.

After the war, living standards changed greatly. More goods were being frozen and goods were being processed all year long. As a result, freezer warehouses have been developed around the lift truck. Frozen foods require air circulation and store easily in storage racks.

Another big boost to lift truck usage was the adoption of the unit-load concept in the food industry. The 40-by-48-in. flush pallet became fairly standard. Through unit loads, a loaded freight car could be emptied in 25 to 30 minutes, compared to as long as 16 hours when unloaded manually.

Some of the guide lines developed specified at least a 20-ft high ceil-

Applications: Problem Solving with Lift Trucks | 221

FIG. 15-5. In addition to transporting goods, the lift truck can serve as a materials elevator. Inside a building, loads can be lifted from one floor to another or to a mezzanine.

ing. Single story construction allowed larger bays which, in turn, facilitated storage layout (fewer columns to worry about).

Floors must be strong enough to support the loaded lift trucks. In addition to strength, the floor must also be designed to resist buckling

222 | Lift Trucks

FIG. 15-6. Low-mast truck allows load handling in a low-ceiling area. Courtesy of Exide

caused by frost. Most warehouse designers either provide ductwork under the floor to allow air circulation or place heating coils in the ground to prevent the buildup of frost.

Lighting is most important and emergency lights should be provided in case of power failures. Lighting in active areas of the warehouse should be a minimum of 20 footcandles and in inactive areas a minimum of 10 footcandles. Exact lighting needs will depend on the conditions encountered, such as colors, outside lighting available, etc, but 20 footcandles allows one to read his own handwriting.

Freezer doors are custom built but they should be built high enough to accommodate any stacking equipment and yet be held to a reasonable minimum to prevent excessive loss of refrigerated air. Air locks and powered door openers are often used in freezer warehouses.

A recent trend in freezer warehouses has been to increase ceiling heights to 30 ft and higher. This trend naturally prohibits the use of a lift truck inside the freezer; however, lift trucks are still required for the shipping docks.

Variety

When selecting a handling device, it is important to consider other handling methods. Is there a better way to handle an object? How does it coordinate with other handling operations? Do suppliers or receivers use the same equipment?

The lift truck is without peer when it comes to options. No matter how a load is shipped into the average plant, there is usually some form of lift truck that can be obtained for unloading and moving the load.

Existing building layouts are easily worked around as in the case of narrow aisle storage. As shown in fig. 15-7, the swiveling mast allows the storing of long objects in tight quarters. If floor loadings are a problem,

FIG. 15-7. Long objects can be stored in tight quarters by selecting the proper truck. Courtesy of Drexel Industries, Inc.

an outrigger truck can usually be found that is light enough to serve the purpose. And, if capacity is a problem, a truck one size larger can be used so that two loads can be handled at one time. Regardless of the requirement made of the load — rotated, upended, etc — a device can be added to the front of the truck to meet this requirement.

Economics

In the long run, economics will dictate the selection of equipment. This is one area where the lift truck has an advantage over all other forms of handling equipment. Its versatility and flexibility, coupled with the variety of units that are available, plus the renting, leasing, or buying plans, make a lift truck practical in all kinds of operations. Almost any other handling method dictates a higher investment and does not offer the option of being changed if the operation does not prove feasible.

Future Trends

Some time ago, the lift truck appeared to have reached its plateau of development. However, as other technologies advanced, it was found that many new methods could be applied to the lift truck. Electronic developments have increased the operating life of the electric truck; LP gas, catalytic exhaust, and automatic transmissions have eliminated many of the air pollution problems associated with the gas-powered truck.

Users are more knowledgable and suppliers know it. No longer are trucks sold as individual items — they are sold as part of a handling system and they are expected to perform economically.

A further advantage of the lift truck industry is that the industry has kept pace with automation. The guided tow tractors and, more recently, the guided lift trucks, offer the industrial user automation with "off-the-shelf" equipment at "off-the-shelf" prices.

Appendix

Present worth and capital recovery factors
(for 10% annual compound interest)

Number of years	Present worth	Capital recovery
1	0.90909	1.10000
2	0.82645	0.57619
3	0.75131	0.40211
4	0.68301	0.31547
5	0.62092	0.26380
6	0.56447	0.22961
7	0.51316	0.20541
8	0.46651	0.18744
9	0.42410	0.17364
10	0.38554	0.16275
11	0.35049	0.15396
12	0.31863	0.14676
13	0.28966	0.14078
14	0.26333	0.13575
15	0.23939	0.13147

Appendix

Formulae and Conversion Factors

1. Travel times in minutes $= \dfrac{\text{distance in feet}}{\text{MPH} \times 88}$

2. Engine torque in foot pounds $= \dfrac{5252 \times \text{horsepower}}{\text{RPM}}$

3. Engine speed (RPM) $= \dfrac{168 \times R \times \text{MPH}}{r}$

    where r = rolling radius of tire in inches
    R = overall gear reduction including both axle and transmission

4. S.A.E. horsepower $= \dfrac{N \times B^2}{2.5}$

    where N = number of cylinders
    B = bore of cylinders in inches

5. Current $= \dfrac{\text{electromotive force}}{\text{resistance}}$ or $I = \dfrac{E}{R}$

    where E = volts, I = amperes, and R = ohms

6. Power (watts) = volts × amperes or, $W = EI$
7. 1 horsepower (hp) = 33,000 ft lbs per min
8. 1 horsepower = 745.7 watts
9. 1 kilowatt = 1,000 watts
10. 1 kilowatt = 1.341 horsepower
11. 1 mile = 1.609 kilometer
12. 1 meter = 39.37 inches
13. 1 gallon = 231 cubic inches
14. 1 cubic foot = 7.48 gallons
15. Specific gravity indicates the ratio of a given volume of material to the weight of an equal volume of water. For example: If a material has a specific gravity of 1.5, multiply this by 62.4 (which is the weight of 1 cu ft of water) to obtain the weight of the material. Material = 1.5 × 62.4 = 93.6 lbs per cu ft.

Appendix

Lift Truck Manufacturers

Air Technical Industries
7501 Clover Avenue
Mentor, Ohio 44060

Allis-Chalmers Materials Handling Div.
21800 S. Cicero Ave.
Matteson, Ill. 60443

Ampulco, The American Pulley Co., Div. of Universal Amer. Corp.
4200 Wissahickon Ave.
Philadelphia, Pa. 19129

Baker Division, Otis Material Handling
8000 Baker Ave.
Cleveland, Ohio 44102

Baker-York
981 South 8th St.
West Memphus, Arkansas 72301

Barrett-Cravens Co.
630 Dundee Road
Northbrook, Illinois 60062

Big Joe Mfg. Co.
7225 N. Kostner Ave.
Chicago, Ill. 60646

Blue Giant Equipment Corp.
2323 Kennmore Avenue
Buffalo, N.Y. 14207

BT Lift Products from Sweden Inc.
187 Mill Lane
Mountainside, N.J. 07092

Camet Industries Co.
500 Lincoln St.
Allston, Mass. 02134

Champ Corp.
2500 N. Rosemead
El Monte, Calif. 91733

Clark Equipment, Industrial Truck Div.
Battle Creek, Mich. 49016

The Colson Company
39 South LaSalle St.
Chicago, Illinois 60603

Crown Controls Corp., MH Div.
Moder Avenue
New Breman, Ohio 45822

Crutcher Resources Corp.
3312 Marquart
Houston, Texas 77027

Appendix

Datsun Forklifts, International Equipment Co.
Mayfield, Kentucky 42066

Drexel Dynamics Corp.
Horsham, Pa. 19044

Eaton Corporation
Suite 800
Foxcroft Sq. Pavillion
Jenkintown, Pa. 19046

The Elwell-Parker Electric Co.
4205 St. Clair St.
Cleveland, Ohio 44103

Erickson Corp.
211 St. Anthony Blvd. N.E.
Minneapolis, Minn. 55418

E-Z-Way Systems, Div. Industrial-Automotive, Inc.
595 W. Church St.
Newark, Ohio

Harlo Products Corp.
4210 Ferry St.
Grandville, Mich.

Hyster Company
P.O. Box 2902
Portland, Oregon 97208

Komatsu Ltd.
Akasaka, Tokyo, Japan

R. G. LeTourneau, Inc.
P.O. Box 2307
Longview, Texas 75601

Lewis-Shepard Co., a subsidiary of Hyster Company
125 Walnut St.
Watertown, Mass. 02172

Lift Trucks Inc.
Cincinnati, Ohio 45214

Lo-Lift Mfg. Co.
4832 Ridge Rd.
Cleveland, Ohio 44144

Matbro Sales Ltd.
British Aircraft Corp. USA, Inc.
Intratec Div.
399 Jefferson Davis Highway
Arlington, Va. 22202

Mitsubishi International Corp.
277 Park Ave.
New York, N.Y. 10017

Moto-Truc
12401 Taft Ave.
Cleveland, Ohio 44108

Multi-Pallet Fork Lifts, Inc.
5237 American Avenue
Modesto, California 95350

North American Man. Co.
Box 1917
Sioux City, Iowa 51102

Pettibone Mercury Corp., a subsidiary of Pettibone Mulliken Corp.
4700 W. Davision St.
Chicago, Illinois 60651

The Prime Mover Co.
Muscatine, Iowa 52761

The Raymond Corp.
3318 Madison St.
Greene, N.Y. 13778

Revolator Co.
2008 86th St.
North Bergen, N.J. 07047

Silent Hoist and Crane Co., Inc.
841-877 63rd St.
Brooklyn, N.Y. 11220

Standard Mfg. Co., Inc.
4012 West Illinois Ave.
Dallas, Texas

Appendix | 229

Stratton Equip. Div. of Hunter Mfg. Co.
30525 Aurora Rd.
Cleveland, Ohio 44139

Taylor Machine Works, Inc.
Louisville, Mississippi

Towmotor Corp., a subsidiary of Caterpillar Tractor Co.
16100 Euclid Avenue
Cleveland, Ohio 44112

Toyo Umpanki Co.
U.S. Materials Handling Inc.
2295 Administration Drive
St. Louis, Mo. 63141

Toyoto Motor Sales U.S.A., Inc.
Industrial Equip. Div.
Box 2991
Torrance, Calif. 90509

Truck-man Div. of Knickerbocker Co.
603 Liberty
Jackson, Mich. 49203

W.S. Tyler Co.
8200 Tyler Blvd.
Mentor, Ohio 44060

FWD Wagner Inc.
P.O. Box 20044-T
Portland, Oregon 97220

Weld-Built Products, West Bend Equipment Corp.
West Bend, Wisconsin 53095

White Industrial Div., White Motor Corp.
301 Ninth Avenue So.
Hopkins, Minn. 55343

Index

Analysis
    bar chart, 136-37, fig. 9-7. 139
    cost, 91-92
    equipment, 83
    financial, 83-84, 87
    layout, 125-27, 129
    lift truck, tab. 4-1. 41
    materials handling, 4
Ampere-hours, 78
Apron space, 187, tab. 13-1. 191
Accessories, rack, 177-78
Attachments (forks), 45-47
    clamping 144-45
    maintenance of, 119
    nonpowered, 140-42
    powered, 142-46, 148
    rotator (inverter), 143-44
    sideshifter, 142
    styles of, figs. 10-7. 147, 10-8. 147, 14-5. 206, 14-6. 207, 14-7. 208
    towing, 211-12
    vacuum, 145-46
Automatically guided tractors, 22, fig. 3-3. 24

Bar chart, analysis using, 136-37, fig. 9-7. 139
Batteries, 71-72
    capacity of, in ampere-hours, 78
    charging methods for, 80-82
    fuel cell, 80
    in gas trucks, 111-12
    lead-acid, 78-81
    maintenance of, 114-16
    nickel-iron-alkaline, 78, 81
    sizes of, 80
Battery charger, 80
Berths, dock, 187-89

Bevel gears, 70
Body, counterbalanced truck, 10
Brake linings, 51-52
Brakes
    "deadman's," 21, 52
    disc, 51-52
    parking, 50
    self-energizing, 50-51
    shoe, 51
Braking
    dynamic, 52
    regenerative, 52
Building
    maintenance, 121
    requirements, for yard handling, 201
    and yard storage, 199-201
Bumpers, dock, 192

Cantilever (Xmas tree) racks, 183
Capital recovery factors, 225
Carbon pile, 76
Charger, battery, 76
Clamping attachments, 144-45
Clutch, 65-66
Communications, 191
Computer
    and materials flow, 132
    and replacement schedule, 89, tab. 6-2. 91
Construction sites, 208, 210-11
Containerized handling, 205-7
Containers
    sizes of, fig. 11-17. 170
    styles of, 155
    trends in, 171
Conversion factors and formulae, 226
Corrugated boards, 163-65
Cost analysis, 91-92

231

Counterbalanced trucks
  body of, 10
  components of, fig. 4-1. 42
  description of, 6-7, 27
  lifting mechanism of, 10, 12
  and racks, 174
  stand-up, 27-28
Counterweight, 12, fig. 2-8. 14
Crane trucks, 202-3
"Cube"
  in rack storage, 1, 179, 182-83
  and narrow aisle trucks, 27
Cylinders, packaging, 165-67

"Deadman's" brakes, 21, 52
Declining balance depreciation, 86
Depreciation
  importance of, 86-87
  rates, tab. 6-1. 87
  declining balance, 86
  straight-line, 85
  sum-of-the-years' digit, 86
Diesel engines, 63-64, 109
Disc brakes, 51-52
Docks
  apron space for, 187, tab. 13-1. 191
  berths on, 187-89
  and building style, 187
  bumpers for, 192
  and communications system, 191
  design of, 186-87, 189-90
  dimensions of, 189
  examples of, figs. 13-2. 189, 13-3. 190
  height of, 192-93
  leveling devices for, 193-94
  location of, 186-87
  outdoor considerations in, 189-90
  safety features for, 190-92
Dockboards, 193-94
Drawbar pull, 14, 16, tab. 2-3. 18
Drive-in racks, 179-83
Drive-through racks, 179-83
Driver, lift truck
  designing for, 53-55, 57
  inefficiency, 97
  safety rules for, 99, 101-2
  training programs for, 98-99
  training of, 96
Drums, 165, fig. 11-14. 167

Electric trucks, 71-73, 75-78
  carbon pile in, 76
  criteria for selection of, 59-60
  gradability, 14
  lifting mechanism, 12
  maintenance of, 114-16, 118-19
  solid state devices in, 76
  stepped resistance control in, 76-77
  torque rating of, 75
Engines
  diesel, 63-64
  internal combustion, 60-61, 106
  liquified petroleum (LP), 62-63, 106-7
Equipment
  analysis, 83
  comparison table, tab. 6-3. 95
  financing, 83-95
  leasing of, 92
  ownership of, 93
  rebuilt, 94
  rental of, 92-93

Finance and equipment ownership, 85-87, 93
Financial analysis, 83-84
  Materials and Allied Products Institute (MAPI) method of, 87-88
Financing, methods of, 83-84
First-in, first-out (FIFO) storage, 179
Flow chart, 127, 129
Flow racks, 177, 179
Formulae and conversion factors, 226
Free lift, 46-47, fig. 2-6. 12
Forks
  styles of, 46-48
  multidirectional, 35-36
  See also Attachments
Fuel cell, 80

Gas trucks, 58-71
  clutch for, 65-66
  counterweight in, 12, fig. 2-8. 14
  gears for, 70-72
  gradability of, 14
  lifting mechanism of, 12
  maintenance of, 60-63, 112-14
  mufflers of, 61-63
  power train of, 58-59
  transmission of, 58-59
Gauges, 54, fig. 4-14. 56
Gears
  bevel, 70
  planetary, 71
  spider, 70-71
Gear pump, 44
Gear train, 65
Gradability, 12, 14-15
Grade
  calculation of, 12, fig. 2-10. 15
  clearance, 14, fig. 2-11. 17

Index | 233

percentage, conversion table for, tabs. 2-1, 2-2. 16
Guidance systems, 22, 23, 25

Honeycombing, 183-84
Horsepower rating, 75
Hydraulic systems, 41-47
　maintenance of, 119-20
Hydrometer, 78, 116-17

Identification
　in packaging, 160-61
　in yard storage, 197
Internal combusion engines, 60-61, 106

Last-in, first-out (LIFO) storage, 179
Layout, plant, 123-39
　agenda for, 125
　analysis of, 125, 217, 129
　and communications, 131-32
　and departments, 123-25
　examples of, fig. 9-2. 128
　and flow chart, 127, 129
　and process flow, 127
　and product flow, 127
　records, 125, 127, 131
　storage, 129-31
　and storage racks, 174, 176
　and warehouse, fig. 13-1. 188
Lead-acid batteries, 78-81
Leased equipment, 94
Leveling devices, 193-94
Load capacity, 7-8
Load center, 7-8
Load center, fig. 11-8. 159
Lift height, 10
Lift truck manufacturers, 227-29
Lift trucks
　analysis system for, tab. 4-1. 41
　and cold storage, 220-23
　design of, 40-47, 49-55, 57
　economics of, 224
　flexibility of, 220
　future trends in, 224
　history of, 1-3
　operating characteristics of, 217-19
　problem solving with, 213-24
　terminology of, 6-20
　types of, 21-38
　versatility of, 213-16
　See also Electric trucks, Gas Trucks, Trucks
Lifting mechanism
　characteristics of, 9-10
　on counterbalanced trucks, 10, 12
　on electric trucks, 12
　on gas trucks, 12
　See also Masts
Liquified petroleum (LP) engines, 62-63, 106-7

Maintenance, 120-21
　of attachments, 119
　of brakes, 50
　of buildings, 121
　combination program for, 105
　contract, 105
　and cost analysis, 92
　of diesel engines, 109
　and downtime, 104-5
　electric truck, 114-16, 118-19
　　battery, 114-16, 118
　gas truck, 107-14
　　battery, 111-12
　　cooling system, 111
　　electrical system, 109-11
　　fuel system, 109
　of hydraulic systems, 119-20
　in-house, 105
　of lubrication, 107-8
　records, 105-6
　of rental equipment, 93
　and safety checks, 105-7
　service procedures, 112
　and sludge, 109
　summary, for electrick trucks, 118-19
　summary, for gas trucks, 112-14
　and yard storage, 199-201
Machinery and Allied Products Institute (MAPI), 87-88
Magnetic guidance system, 23-24
Mast
　articulated, 35
　carriage assembly on, fig. 2-4. 9
　fixed, 9-10
　four-stage, fig. 2-5. 11
　selection of, 45-46
　two-stage, 10
　telescoping, 10
　See also Attachments
Materials handling
　analysis of problems in, 4
　history of, 1-3
　reasons for, 3-4
　and simulation, 135-36
　styles of, fig. 14-8. 209
Mezzanine, 185
Motor
　compound, 75-76
　propulsion, 75

pump drive, 75
shunt, 75
Muffler, 61-62

Narrow aisle trucks, 27-28
  and "cube," 27
Network diagram, 136-38
Nickel-iron-alkaline batteries, 78, 81
Noncounterbalanced trucks, 7, 9
Nonpowered attachments, 140-42

Occupational Safety and Health Act (1970), 96-97
Optical guidance system, 23-24
Order-picking loads, 176
Order-picking trucks, 36-39
Outrigger trucks, 7, 9, 27-29
Overhead guard, 17-18

Pallet, 168, 150-52, fig. 11-6. 156
  and flow racks, 177, 170
  four-way, 152-53
  loads on, 155, 157
  slave, 152, 155
  in storage racks, 174-75
  styles of, fig. 11-5. 154
  two-way, 152-53
  types of, 152, 155
  and unit loads, 155
Pallet trucks, 25-26
Packaging
  and humidity, 160, fig. 11-10. 162
  identification on, 157-58, 160
  methods of, 157, 160, 163-66, 168, 171
  paper, 157
  shrink film, 157
Paper shipping sacks, 160, 163
Planetary gears, 71-72
Plant layout. *See* Layout, plant
Pollutants, 61-63, 106-7
Power steering, 44-46
Power trains, 58-62. *See also* Electric trucks, Gas trucks
Powered attachments, 142-46, 148
Problem solving, 213-24
Process flow chart, 127, 129
Product flow chart, 127, 129
Program Evaluation Review Technique (PERT), 136-37

Rack accessories, 177-78
Rack clearance, 172-73
Racks
  cantilever (Xmas tree), 183-84
  and counterbalanced trucks, 174

"cube," 179, 182-83
drive-in, 179, 182-83
drive-through, 179, 182-83
and floor loading, 172-73
flow, 177, 179
height of, 176
installed in existing buildings, 172-75
and load characteristics, 174
mezzanine, 185
self-stacking pallet, 185
and sideloaders, 29-32, 174
standard adjustable pallet, 175-77
variety of uses of, 177
Rack bay, 176
  buildings, 184-85
  system, layout of, 172-75
Rating system for trucks, 40-41
Reach trucks
  carriage type, 29-30
  sideloaders, 29-32
Rebuilt equipment, 94
Receiving docks. *See* Docks
Records
  layout, 125, 127, 131
  maintenance, 105-6
Rental equipment, 93-94
Replacement schedule
  computerized, 89
  formula, fig. 6-3. 90
  for lift trucks, tab. 6-2. 91
  policy, 87, 89
Roadable units, 198-99
Rotator (inverter), 143-44
Rubber-tired traveling crane, 202

Safety
  on docks, 190-92
  driver, 99, 101-2
  checks, maintenance, 105-7
Security, 201
Self-energizing brakes, 50-51
Shipping docks. *See* Docks
Shoe type brakes, 51
Shrink film packaging, 157-58
Sideloader, 29-30, fig. 14-4b, 204
  advantages of, 31
  and construction, 208-11
  disadvantages of, 32
  and racks, 174
Sideshifter, 142
Simulation
  in dock design, 189
  and materials handling, 135-36
Skid, 151-52, fig. 11-6, 156. *See also* Pallet
Sludge, 109

Solid state devices, 76-77
Speed
  lift, 19
  travel, 19
Spider gears, 70-71
Stability, 16-17, fig. 11-8. 159
  and stacking pattern, 157
Standard lift truck, fig. 14-4a. 204
Standard steel adjustable rack, 175-77, fig. 12-6. 182
Steering
  articulated, fig. 4-9. 49
  cylinder, fig. 4-4. 45
  power, 44-45
  systems, 47, 49
Stepped resistance control, 76-77
Stillage (skid) trucks, 26-27
Storage methods, 20
Storage yard. *See* Yard storage
Straddle truck, 7, 32-33, 35
Straight line depreciation, 85
Sum-of-the-years' digit depreciation, 86
Supervisor, rules for, 101-2
Swingload, fig. 14-4c. 204
Swiveling mast trucks, 34-36

Terrain, and yard storage, 197-99
Termostat, 111
Tiering, 20
Tilting capability, 10, fig. 2-7. 13
Time studies, 132-34
Time values, 133
Timing, 110
Tipping characteristics, 16-17
Tires
  cushion, 52-53
  maintenance of, 120-21
  pneumatic, 52-54
  with power steering, 45
Towing attachments, 211-12
Torque rating, 75
Torque, dynamic braking, 52
Tractors, tow, 21-25
Trailers, 21-25
Training program for drivers
  classroom, 98
  driving test in, 98-99
  field, 98
  safety rules in, 99, 101-2
Transmission
  automatic, 66-69
  standard, 65, fig. 5-5. 67
Trucks
  crane, 202-3
  counterbalanced. *See* Counterbalanced trucks
  diesel, 63-64. *See also* Gas trucks
  electric, 59-60, 71-72, 75-78, 118-19. *See also* Electric trucks
  gas, 58-71. *See also* Gas trucks
  hybrid, 203, 205
  narrow aisle, 27-28
  noncounterbalanced, 7, 9
  order-picking, 36-38
  outrigger, 7, 9, 27-29
  pallet, 25-26
  reach, 29-32
  sideloader, 29-32
  sideshifter, 142
  standard, 14-4a, 204
  stillage (skid), 26-27
  straddle, 7, 32-33, 35
  swiveling mast, 34-36
  walkie, 53-55
Turning radius
  inside, 19
  outside, 17, 19
  of trailers, 21-22

Unit load, 149-51
  and outrigger capability, 28
  and pallets, 155
  and storage racks, 176

Vacuum attachments, 145-46

Walkie trucks, 53-55
Warehouses, cold storage, 22-23
Wheels, steering, 49-50
Wood boxes, 165

Yard handling
  and building layout, 199-201
  lighting, 199
  maintenance, 199
  security in, 201
Yard storage
  and identification, 197
  problems of, 195-96
  and product analysis, 196-97
  and terrain, 197-99